Systems Theory and Theology

Systems Theory and Theology
The Living Interplay between Science and Religion

Edited by
MARKUS EKKEHARD LOCKER

PICKWICK *Publications* · Eugene, Oregon

SYSTEMS THEORY AND THEOLOGY
The Living Interplay between Science and Religion

Copyright © 2011 Wipf and Stock Publishers. All rights reserved. Except for brief quotations in critical publications or reviews, no part of this book may be reproduced in any manner without prior written permission from the publisher. Write: Permissions, Wipf and Stock Publishers, 199 W. 8th Ave., Suite 3, Eugene, OR 97401.

Pickwick Publications
An Imprint of Wipf and Stock Publishers
199 W. 8th Ave., Suite 3
Eugene, OR 97401

www.wipfandstock.com

ISBN 13: 978-1-60608-739-8

Cataloguing-in-Publication data:

Systems theory and theology : the living interplay between science and religion / edited by Markus Ekkehard Locker.

xiv + 228 pp. ; 23 cm. Includes bibliographical references and index.

ISBN 13: 978-1-60608-739-8

1. Religion and Science. I. Locker, Markus. II. Title.

BL240.3 L635 2011

Manufactured in the U.S.A.

Contents

List of Illustrations / vi
Preface / vii

1. A Concise Introduction to General and Trans-Classical Systems Theory—*Alfred Locker* / 1
2. Systems Theory and the Axis of Faith: Creation and Evolution—Resurrection and Life Everlasting —*Alfred Locker* / 18
3. Edenology: A Science of Paradise?—*Otto Rössler* / 38
4. Personal Knowledge and the Inner Sciences —*Martin Zwick* / 49
5. Symbolic Structures as Symbols: On the Near Isomorphism of Two Symbolic Structures—*Martin Zwick* / 62
6. Boundary Discourse as Religious Communication —*David Krieger* / 97
7. Disruptions: Systems and Investigations —*Clemens Sedmak* / 127
8. Portal, System, and Sacred Order: America —*Peter Murphy* / 143
9. Perceiving Freedom and Meaning in Nature: Operationalizing Trans-Classical Systems Theory for Converging Scientific and Religious Knowing —*Markus Locker* / 176
10. Scripture and Trans-Science: Parables as "Systems" of the Kingdom—*Markus Locker* / 192

Selected Bibliography / 211
Contributors / 225
Index /

Illustrations

Figures

1.1	Allology—Autology Schema / 9
2.1	The Super-Systems of *First & Last Things* / 37
5.1	Diagram of the Supreme Pole and the Kabbalistic Tree / 64
5.2	Differentiating and Integrating Triads / 85
10.1	The Communicative Activity of the Parable-System / 197
10.2	The Environments of Parable-Systems / 199
10.3	Objective System Observer / 202
10.4	Communicating System Observer / 202
10.5	Trans-classical Systems / 203
10.6	Parable (P) Interaction and Transformation / 206

Tables

5.1	Enumeration Orders of the Five Agents / 74
5.2	The Five Agents and *Sefirot* 4–8 / 76
5.3	The Five Virtues and *Sefirot* 4–8 / 77
5.4	Labels of the Diagram of Chen Tuan / 88

Preface

THE FLOOD OF PUBLICATIONS on the relationship between science and religion proves that not many issues have received as widespread attention as the possible dialogue and relationship between the natural sciences and religion. The overall assessment of the success of many of these attempts remains, however, inconclusive and doubts cannot be dispelled whether a strictly scientific worldview and an understanding of the universe through the eyes of faith can ever settle for a mutually acceptable common denominator. Different from the more common attempts to search the discoveries of advanced sciences and the age-old truths of faith for potential concord and commonalities,[1] the present volume—*Systems Theory and Theology*—actively seeks and engages in an interplay between scientific theory and religion. This effort is rooted in the life-long work of the late Austrian physicist Alfred Locker (1922–2005)[2]—one of the last students of Ludwig von Bertalanffy (1901–1972) the acknowledged founding father of General Systems Theory—whose work in the firm belief that systems theory will at long last emerge as bridge-building meta-theory pursued a twofold aim; on the one hand to advance General Systems Theory into a universal tool for joining and combining insights from science, philosophy, and theology, and on the other hand to illumine ostensibly contradictory philosophical and theological viewpoints on the basis of advanced scientific theory. The gradual realization of the latter goal was initiated through the proposal of Trans-classical Systems Theory in 1972.[3] Many years later the editor[4]

---

1. The paper of George L. Murphy, "Does the Trinity Play Dice?," *Perspectives on Science and Christian Faith* 51 (March 1999): 18-25 represents the currently very popular view to liken quantum phenomena with supernatural realities.

2. Jung, "Alfred Locker: An Obituary," 1665–66; Glanville, "Learning from Locker," 223–27; Pichler, "Alfred Locker Im Gedenken," 59; Locker, "Glimpses of Truth," 103–5; "Obituary," 571–75.

3. Alfred Locker, "Ontological Foundations," 537–71.

4. Markus Locker is the son of Alfred Locker.

of this volume—himself a theologian with then amateurish interest in systems theory—came to realize that Locker's systems theoretical work has significant implications for the interpretation of religious texts and theology in general. After initial successful attempts to explore possible interaction between systems theory and theology in academic conferences[5] and publications,[6] the idea was born to explore further possibilities in employing systems thinking and systems theory to illuminate religious content in depth. In 2003 Alfred Locker accepted the invitation to pursue this goal in preparing for a collection of essays in collaboration with close friends and colleagues. It became the task of the editor to invite, over the next few years, further contributions to this volume. At long last—however five years after the death of Alfred Locker—enough papers could be compiled for what ultimately became an assortment of contributions from scholars throughout the whole world.

The first essay of this anthology is one of the last papers by Alfred Locker, written shortly before his death in 2005. Entitled *A Concise Introduction to General and Trans-Classical Systems Theory*, it remained fragmentary and in part unfinished, bequeathing the rewarding task to the editor to complete and translate the manuscript faithful to its author's intentions. Locker's essay is deliberately complex, densely written, and replete with enduring thoughts. Yet it undoubtedly represents his ideas spanning over more than thirty years *in nuce*. Heavily influenced in style and tenor by the great minds that shaped his work—among many others, Johann Georg Haman, Goethe, and Novalis—Locker's Trans-Classical Systems Theory (TCST) envisages a holistic systems theoretical approach that, at first, exposes the strength and weaknesses of General Systems Theory (GST) in dealing with the *ganzheit* of systems and reality as a whole. The key for a universal systems theory is the otherness (allology) and simultaneous sameness (autology) of a system to its observer due to the fact that even before systems are delimited or conceived the theory behind the system, or systems theory, must already exist as concept or idea. The philosopher's stone of systems theory is found in the sense of balance in simultaneously viewing systems as subject *analogs* and *allologs*, hereby equally recognizing the systemic nature

---

5. The support of the IIAS and its president Prof. George Lasker must be mentioned here. Regrettably Prof. Lasker—due to health reasons—was not able to complete his contribution to this volume.

6. Locker, "System-Theorie," 16-22; "Das Buch der Offenbarung," 3-10.

of the knowing subject. Full access to any system is provided neither by an *exo-*, or an *endo-*perspective, but by immersing the human access system into the system of concern.

Consequently, typical systems observation is bound to reveal only a very limited part of the whole system and has to be augmented by approaches like perception and participation, eventually leading to a synthesis of all viewpoints converging at a definitive view tantamount to "beholding" a system. The combination of viewpoints, however, results in the emergence of paradoxes and contradictions in the system studied that cannot be satisfactorily addressed by means of classical logic and mathematics as employed by General Systems Theory. Trans-Classical Systems Theory proposes that a holistic study of systems requires the use of non-classical means of knowledge, much akin to Romantic *poetization* and religious *contemplation*. One the basis of the inseparable unity of presuppositional properties and objective features of a system, TCST continues to argue that the nature of a system cannot be inferred by observing and measuring its functions, or their changing in time, i.e., evolution, but progressively reveals itself on account of the corresponding transformation and *transfiguration* of the subject-system. Locker concludes that systems theory is a two-edged sword, on the one hand all too readily prepared to restrain reality to arguably limited scientific modes of inquiry, while on the other hand truly able to become conscious of systems *in toto* and "all of reality" in its demonstrably systemic entirety.

This introductory paper is followed by Alfred Locker's essay *Systems Theory and the Axis of Faith: Creation and Evolution—Resurrection and Life Everlasting*. Grounded in Trans-Classical Systems Theory Locker undertakes to study the properties of the systems of "First Things" and "Last Things" as portrayed by science, philosophy, and theology. Locker shows that approached with Trans-Classical Systems Theory the alleged incommensurable properties of these systems can be viewed in complementary unity. This systemic viewpoint is however replete with paradoxes that need to be addressed with recourse to the human access system to the systems of "origin" and "end." Locker continues to stress the time-transcending existential and constitutive relationship of subject and system reaching the conclusion that a unifying understanding of reality as offered by systems theory not only provides for the critical reconciliation of worldviews, but serves to strengthen Man's relationship with transcendence.

The paper of Otto E. Rössler—*Edenology: A Science of Paradise?*—re-addresses three scientific topics with an eye to the question concerning the status of the problem of the now, the theory of mothering, and Everett's quantum worlds. The two unfamiliar technical terms that are identified are "assignment conditions" and "benevolence theory." The former notion needs to be added to the 320-year-old technical notions of "initial conditions" and "laws." The latter revives the forgotten technical term of the "sun of the good." The existence of assignment as a fact of nature calls for a better understanding of its nature, acceptability, modifiability and, most of all, intention.

Martin Zwick's article *Personal Knowledge and Inner Sciences* conceptualizes spiritual disciplines as sciences and uses this conceptualization to probe into the similarities and differences between modern science and religious tradition, and the cultural significance and possible future impact of the new religions. Drawing upon the ideas of Michael Polanyi as building a possible bridge between science and religion, it is proposed that his ideas are likewise relevant to Eastern and non-mainstream Western religions. Imagining science as a spiritual path, or gnosis, opens not only genuine challenges to any exclusivist understanding of scientific knowledge apart from wisdom of being, but likewise provides a novel basis for a dialogue between science and religion.

The interpretation of the spiritual disciplines as inner sciences leads to a critique of science and a new conception of its possibilities, and might even contribute to a strengthening and purification of religious practice. Since, however, here are many major differences between the "inner" and "outer" sciences the metaphor is admittedly limited, and if taken too literally, it will assuredly obscure more than it illuminates.

Symbolic Structures as Symbols: On the Near Isomorphism of Two Symbolic Structures[7] by Martin Zwick explains that many symbolic structures used in religious and philosophical traditions are visibly composed of elements and relations between elements. Similarities between such structures can be described by the systems theoretic idea of "isomorphism." This paper demonstrates the existence of a near isomorphism between two symbolic structures: the Diagram of the Supreme Pole of Sung Neo-Confucianism and the Kabbalistic Tree of medieval Jewish mysticism. The similarities of these two symbols in form and meaning

---

7. Published in shorter form in *Religion East and West, the Journal of the Institute for World Religions* 9 (October 2009) 67–87.

are remarkable in the light of the many differences between Chinese and Judaic thought. Intercultural influence might account for these similarities, but a more parsimonious explanation would attribute them to the ubiquitousness of ideas about hierarchy, polarity, and macrocosm-microcosm parallelism.

David Krieger, a well-published author on systems theory, argues in *Boundary Discourse as Religious Communication* that the way in which a system of meaning differentiates itself from meaninglessness, chaos, and disorder constitutes the function of religion. Religion is important not because people believe in gods or revelation of some kind, but because boundaries between self and other, meaning and meaninglessness must be drawn in every instance of systemic and subsystemic order. Drawing these boundaries requires a specific form of communication, a "boundary" discourse that may be characterized as religious communication. The pragmatics of boundary discourse includes proclamation, narrative repetition, ritual representation, temporal orientation towards founding events, and inclusion/exclusion. In general, it may be said that concepts of the absolute are nothing but the necessary self-reference of the meaning system, and are constructed by communication according to specific pragmatic rules and conditions.

Clemens Sedmak's broad experience as theologian, philosopher, and epistemologist causes him to raise the fundamental question: *Do we need Systems in Theology*? He poses this question in three ways: 1) Theological work in our times has to strive towards "local theologies" to answer the question of a "fulfilled life." Is the systems theory approach compatible with these demands? 2) Christianity gives the impression that Jesus had never developed a theological "system." Romano Guardini pointed out that the root of the Christian religion is a person, not a system of doctrines. Is systems theory, therefore, adequate to grasp the concerns of the Christian religion? 3) Doing theology "as if people mattered" asks for a clear language and for respect for differences ("local theologies are messy"). Is systems theory capable of being developed in such a way that these requirements of clear language and respect for differences can be met? Sedmak concludes that there is nothing wrong with theological systems as long as they are disruptable and open to genuine theological reflection.

Peter Murphy's article *Portal, System and Sacred Order: America* adapts Bertalanffy's notion of open systems to show how religion, science, art, and philosophy develop together as the system of civilization. Civilization as a system creates meaning by differentiating between order and chaos. The open system of civilization, though, is paradoxical. It creates clear boundaries between meaning and meaninglessness—by separating order and chaos. But it does this without causing the closure of meaning. This is because systems of civilization are open to their environment. Religion, science, art, and philosophy advance by exchanges with each other and with their environment in the broader sense. They import and export meaning. Theology emerges from the exchange of religion and philosophy, as philosophy does from the exchange of science and art. Murphy's essay explores how the import and export of meaning is organized around "ecumenes"—the portal-like cores of civilizations in history. He discusses the evolution of these "ecumenes" from the standpoint of the philosophy of history.

Markus Locker's *Perceiving Freedom and Meaning in Nature: Operationalizing Trans-Classical Systems Theory for Converging Scientific and Religious Knowing*[8] focuses upon the question whether nature's processes pursue a given goal or emerge through self-generated occurrences, by pointing to the fact that both science and religion view nature and all creatures beside the human person as irrational objects. Maintaining this epistemological caesura between the knowing subject and the known object will indeed not allow for any other conclusion than that intelligible developments in nature are either caused by God or are the result of accidental events.

In the search for a third view that is acceptable to both science and theology, this essay argues that Trans-Classical Systems Theory can provide the meta-theoretical concept through which subjective and objective observations can be brought together. Trans-Classical Systems Theory joins the knowing subject and the object under observation in conceiving of the observer/designer as an access system, whose properties (in similarity and difference) are likewise found in the system of observation. Markus Locker arrives at an understanding of natural processes and occurrences in which the principal property of the human access system that is imperative for understanding its own self within the natural world (i.e., freedom) can be introduced to an understanding of the essence or self of the world. This view, however, will not confuse the

---

8. First published on May 24, 2007 in the online journal, *Global Spiral* (8/2) available at http://www.metanexus.net/Magazine/tabid/68/id/10040/Default.aspx.

natural world with full moral consciousness, but foremost suggest that free natural processes stand in an essential relationship to human persons, thereby obtaining meaning beyond that of being simply designed, or occurring by chance.

It is at this point where science and religion do not have to continue to argue which interpretation of nature obtains more plausibility, but jointly can pursue the search for nature's true purpose.

In *Scripture and Trans-Science: Parables as "Systems" of the Kingdom*[9] Markus Locker argues that after nearly a century of critical parable research it still remains an enigma why some of the most remarkable of these central stories of Jesus continue to cause disbelief in the attentive reader; Why is a wedding guest found without proper garment thrown into darkness (Matt 20:13), and an anxious servant unable to multiply his single talent handed out an overly severe punished (Matt 25:29)? Whereas established scholarship seeks for ways to dispel these facts as moral metaphors that must not be taken literally, Locker believes that the apparent incongruities between the text and the "good news" are real and must be addressed by recognizing that parables are foremost speech systems. By studying the parables under the perspective of systems theory, what appears to be irreducible paradoxes within the plot, points to the need to seek personal interaction and active communication with the system of concern. The joining of exegesis and systems theory hereby proposed offers new ways of reading the parables of Jesus.

The volume *Systems Theory and Theology* can ultimately claim to show noteworthy examples of a truly fruitful collaboration and conjoining of scientific systems thinking and religious conceptions. Systems theory not only aids theology, and correspondingly philosophy, in placing the human recipient of revelation back into the systems of faith language, but also enables and promotes religious discourse and religious, thus inter-religious, communication. If systems theory, in sustaining theology's search for clarity of language, seeks to address fundamental human questions in view of improving the human condition, then at the same time the fundamental gap between nature and God can be bridged, for example by studying and establishing models of autopoietic ecumene among and between civilizations. Even life everlasting and the paradise become accessible to theological systems thinking if systems of theology

---

9. Originally published as "Scripture and *Meta*-Science: Parables as 'Systems' of the Kingdom." *The Loyola Schools Review. School of Humanities* IV (2005) 59–83.

are kept sufficiently open. Systems theory supports biblical studies in its quest of authentic communication as well as relates social systems to the core message of the gospel, i.e., charity. In the end the proposal of a lasting embrace and interplay of science and theology can only be seen as encouragement for continuing the dialogue and conversation of *Systems Theory and Theology* joining the world of science and faith.

1

A Concise Introduction to General and Trans-Classical Systems Theory

ALFRED LOCKER (1922–2005)

(translated and completed by Markus Locker)

ABSTRACT

This succinct presentation exposes the strengths and weaknesses of General Systems Theory (GST) in dealing with reality as a whole while remaining firmly rooted in mathematically based methods. More fully recognizing the existing correspondence between any system and the observing subject, it is argued that on account of a system's autology and allology in view of its observer, full access to any system is only granted through simultaneous observation, perception, and participation in the system of concern. Hereby emerging paradoxes require a non-classical treatment provided by the premises of Trans-Classical Systems Theory (TCST). TCST recognizes the inseparable unity of presuppositional (p) properties and objective (o) features of a system, assuming that the whole or gestalt of a system cannot be inferred by simply observing and measuring its functions, or their changes (evolution), but gradually reveals itself through the transformation a system and at the same time the human observer undergo in the process of observation. Systems Theory, as general approach to conceive of reality in a non-deterministic view, thus remains an ambivalent tool; on the one hand prone to restrain reality to arguably limited scientific modes of inquiry, while on the other hand truly able to become conscious of all of reality in its demonstrably systemic wholeness.

A GLIMPSE AT THE HISTORY OF SYSTEMS THEORY

BY AND LARGE CONVENTIONAL scientific thinking retains the maxim that in order to gain true knowledge of material objects and phenomena, all things must first be dissevered and broken up into tractable elements. Subsequent to thorough examination and quantification, these elements can allegedly be re-united with the secondary aim of rendering the originally given subject matter wholly accessible. In general, however, it must be assumed that any method embodying this very scheme is fated to miss this target altogether since the attempt of reconstructing reality after its fragmentation without fail leaves behind an untraceable surplus or remainder. This knowledge existed already in antiquity, commonly known as the Aristotelian dictum that the whole is larger than the sum of its parts, but with the advent of modern science it has been progressively ignored. Science, increasingly enamored of its alleged achievements, never came to realize that its so-called breakthroughs were bought at the costly price of reductionism, i.e., subjugating subject matter (or even worse: subduing it) to states and conditions that allows it to fit the methods of quantitative assessment.

At the turn of the nineteenth to the twentieth century, leading figures in the sciences became increasingly discontented with the persistent fragmentation and deconstruction of reality. The fact that empirical methods captured the entirety of phenomena only to a very small degree, or failed to explain their totality altogether, could not be overlook any longer. Thus, a crisis arose which brought about a gradual shift from an analytical to a synthetic view of reality. The hereby ensuing quest to overcome empirical reductionism resulted in the development and formulation of meta-sciences, labeled as theories of *ganzheit* (wholeness), *gestalt* (form), and system respectively, that remain intimately connected with the names of their proponents, like for example, Othmar Spann (1878–1950), Christian von Ehrenfels (1859–1932) and Ludwig von Bertalanffy (1901–1972), who were all natives of Austria. The most recent meta-science was fleshed out by Ludwig von Bertalanffy and became known as General Systems Theory (GST).[1] Over time, each meta-

---

1. For the historical development of GST, cf. Ludwig von Bertalanffy, "The History and Status of General Systems Theory," and K. Müller, *Allgemeine Systemtheorie*. Important forrunners of GST are G. W. Leibniz with his *Monadologie* (1714), and the German scholar Johann Heinrich Lambert (1728–1777), *Fragmente einer Systematologie* (1787) and *Theorie des Systems* (1782), where the latter attempted to provide a classification of

science underwent developments of different form and pace, such that nowadays systems-theory, on account of its general usefulness, is by far the best known theory, taking center stage in the public consciousness.[2] Throughout the past century, systems-theory and the simultaneously originating science of Cybernetics increasingly dealt with the age-old philosophical problem of subjectivity and the epistemic role of the observer of phenomena, while at the same time progressively addressing the issue of wholeness.

Yet in spite of their shared opposition to scientific reductionism, the methodologies of the two sister-sciences, GST and Cybernetics—though with notable exemptions, like for example Gotthard Günther[3]—continue to adhere to the classical principles of logic and rational consistency, foremost expressed in their avoidance (or deliberate elimination) of ambiguity and paradox. In that sense, GST and Cybernetics remain ostensibly artificial and restricted in handling reality in its assumed entirety. Reality—as perhaps distinguished from excerpts thereof readily accessible and measurable by individual scientific disciplines—by no means is, and can be grasped, entirely free from contradictions and paradoxes. At the same time as a genuine meta-science must be capable of recognizing reality appropriately (i.e., in its entirety), and wield some correcting influences upon it in the event the system is subjected to disturbances, it likewise must be competent to deal with paradoxes innate to all reality. Along these lines, the turn of the twentieth to the present century may be characterized as the transition from a classical to a trans-classical view of reality. Trans-classical thinking, as proposed by this author, will fortify and enhance GST, not only by way of tolerating paradoxes, but by supplying the epistemic capacity to fruitfully employ paradoxes in accessing and comprehending reality. The newly emerging meta-science may thus sensibly be called Trans-Classical Systems Theory (TCST).[4]

---

systems according to their binding forces, but also postulated the need for constraints in order to delimit the concept of system.

2. Lazlo, ed. *The Relevance of General Systems Theory*.

3. Günther, *Die Philosophische Idee*, 24–30.

4. The term *non*-classical has been introduced in the advent of Quantum Mechanics; the term *trans*-transclassical has first been employed by the German logician Gotthard Günther in his attempt to construct a non-Aristotelian Logic which finally resulted as the operationalization of Hegelian dialectic. Cf. R. Glanville, "A Note on Knowing."

TOWARDS A DEFINITION OF SYSTEM AND SYSTEMS THEORY

Reality as System

Exceeding a purely formal definition of system akin to most mechanistic models, von Bertalanffy engendered the modern conception of system on the basis of biology. A system, according to von Bertalanffy, is a complex of entities relating mutually and self-referentially to one another and to the whole of the system, thus providing the latter with the faculty to sustain itself against any disturbing influences threatening the system's existence through stimuli originating from its environment (and/or by irritant processes originating from the system itself). This broad definition, perhaps more aptly considered a circumscription, is self-referential in that the *definiens* equals the *definiendum*. On account of the fact that the herein rendered notion of system is essentially artificial, i.e., conceived and formalized by a human person, it likewise contains and displays traits of its "inventor," which in turn allows for a genuine insight into the system's organization, i.e., its essence or nature.[5] Consequently it may be inferred that each system is a *subject-analog* (i.e., analogous to its observer or designer) and to a certain degree a *substance-analog* (i.e., sharing in the nature of its observer or designer).[6] This fact accounts for the system's self-reference, i.e., the ability to refer to its elements and their relation with one another, and to communicate the difference between itself and its environment, including the observer.[7] What follows is that a formalized system can indeed be viewed as a genuine segment of reality encompassing (or at any rate representing) the latter in a nutshell. Conversely reality itself appears only in the form of differentiation, and thus is to be considered a system.[8] Assuming the veracity of the system-subject correspondence, one significant mode of access to reality is attained that has long been overlooked, viz. the mutual representation of phenomena where everything may be represented by everything else, even in the form of opposites or paradoxes.[9] Systems,

---

5. The organization of living systems is tantamount to the concept of life. Cf. Schöppe, *Theorie paradox*, 142.
6. Ibid., 129.
7. Ibid., 130.
8. Cf. hereto the seminal work of Spencer Brown, *Laws of Form*.
9. Cf. Rothenberg, "Process of Janusian Thinking in Creativity," 195–205.

for that reason, cannot be understood as static—timeless momentary cross-sections of reality—but must be seen as expressing the incessant dynamism of the whole. Trans-Classical Systems Theory is tantamount to a systems-analysis that acknowledges reality's permanent change or transformation.

A Preliminary Systems View

Already an initial and static understanding of systems does reveal significant systems characteristics such as the concepts of self, subject, and substance, but also the notion of wholeness or *gestalt* that must be designated as presuppositional (p),[10] i.e., existing alongside the system's objective (o) properties. Whereas the systems theoretician can encounter these presuppositional properties (that are qualitative and in principle inaccessible to quantification and measurement) through introspection or self-reflection (i.e., by means of obtaining a meta-systems view),[11] only the objective properties of a system are empirically detectable and therefore identical with those usually described in the sciences. The specificity of a system is such that it may be considered on the one hand as the difference between these two kinds of features, and, on the other hand, as the unity of them. The acceptance of both aspects—together with an additional yet fundamental third aspect—viz. that of the immediate, instantaneous origin of the system—compels systems theory to shift its attention incessantly to and fro between these levels, and to hold all three features (in their togetherness forming yet again a system) in suspension hereby demonstrating that a static systems view is in fact unreal and unnatural. This contention underscores some fundamental findings and claims of systems theory. Following the principle of the system's *autology* (AUTo) vis-à-vis its designer—hereby revealing the harmony of p and o properties within the system that is perceived from a meta-level of observation—a system can only be recognized by a system (i.e., something "similar" to itself). The remaining difference and distinctiveness of the system—i.e., the system's *allology* (ALLo)—from its observer and environment is grounded in its identity and unique properties. Systems theory thus stresses the following points:

---

10. The most common p system characteristics—in German designated as Voraussetzungen (V)—of a system can be summed up in the acronym AEIOU coined by Alfred Locker: A, for *autonomy*; E, for *existence*; I, for *individuality* and *identity*; O, for *order*, and U, for *unity*.

11. Cf. Locker, "Der Mensch," 34–42.

1. With regard to the customary view concerning the design of a system and its elements, which by being mutually interconnected with one another constitute the system, it would be erroneously to think that some systems properties are only p, while others are o. Quite the reverse, all systems properties in representing one and the same system are mutually participating in its p and o qualities.

2. As a consequence of this multiplicity of conceptual aspects, it must be admitted that systems are *ubiquitous* (i.e., everywhere or nowhere). That means that in dependence on the chosen perspective, all reality can either be considered obtaining the form of a system, *or* the systemic nature of reality can altogether be denied. This arbitrariness in the identification of systems is only transitory. Eventually within all given phenomena, like for example the human person, some systems will emerge in a permanent recognizable form.

3. Following von Bertalanffy, systems theory in the most general sense aims at bridging the gaps between somewhat unrelated sciences by finding out *isomorphies* in the laws governing particular areas of systems.[12] Hereby a certain "economy principle" emerges if several laws have been found valid for a certain domain. These laws can be rather easily transferred to another domain, provided the similar structure of the latter has been recognized beforehand. For that reason, GST tended to become a somewhat mathematically stretched discipline whose actual receptivity of conflicting viewpoints has been seriously questioned.

The three abovementioned features of systems theory are located on different ontological (and observational) levels, hereby disclosing the possibility of arbitrariness in handling systems and the need to accept exceptionality. Despite acknowledging the concept of systems autology, systems—in the view of GST—are, on the whole, treated as entities essentially different from the idea and notion of the human person. Consequently in the usual undertaking of GST, we—the observers—have exempted ourselves from any (self)-consideration for a "theory of system." This signifies a fundamental deficiency of GST that in the view

---

12. von Bertalanffy, *General System Theory*.

of the author can be redeemed in understanding the human person as the primary, if not the only, "access system" to realty in general.[13]

THE ROLE OF OBSERVER AND PERCEIVER

Internal and External Observer

To begin with, systems theory is in need to maintain the distinction between an external observer (O(e)) and an inner or internal observer (O(i)).[14] The notion of internal observation illustrates particularly well how the concept of observation in general remains subject to the very limitations inherent to any attempt at objectifying given phenomena. Both internal and external observation restrain observation to a single conceptual position, thereby destroying the system's connectedness with other objects and with the ever-changing reality surrounding it. Consequently both modes of observation do not really provide for an insight into objectively given things. An observer will only obtain results that by and large confirm the applied method of observation or measuring, while arguably all those properties of the systems that are not detectable by this inquiry will remain unnoticed. What is more, inner observation can easily be mistaken as factual self-observation leading to views audaciously identifying human persons as machines, or vice-versa, identifying machines as possessing human qualities, like intelligence.[15]

Yet the differences between these two modes of observation are of major importance for systems theory. Inner observation can be obtained by differentiating between o-characteristics of a system (represented by those objects that conform to the applied perspective), i.e., the observations made by the O(e), and their accord with the p-characteristics, provided that the O(i) realizes that he himself bestows these characteristics to the system. This differentiation is usually described as the difference between an *exo*- and an *endo*-view of systems.[16]

---

13. Cf. Locker, "Systems Theory and the Conundrum," 297–317, and here especially p. 300.

14. Rössler, "Endophysics," 154–162, and Locker, "Systems-Theoretical Considerations."

15. Locker, "A.I. and Ethics," 63–68.

16. This is succinctly described in Alfred Locker's paper "'Synolologie' and 'Chaologie,'" 71–101, but acknowledged as a result achieved by other authors. The unusual name has been derived from the concept of Hippocratic Medicine *synolon panton*: All (in the body) is related to all.

Internal observation, however, exerts the same influence on the examined object as external observation does. Both modes of observation continue to alter the object according to the interests of the observer. The suggested way out of this vicious cycle is self-observation. Self-observation—somewhat identical with self-reflection—does not subjugate the object, but influences the acting, i.e., the observing, subject. By means of self-reflection the reflecting subject is re-directed towards itself with the effect of a gradual liberation from subjectivity. Through continuous reflection, perhaps identifiable with the concept of meditation, the subject might finally reach a state of dissipation. The result thereof can be understood as the liberation from the subjectivity of the self.[17]

Within the confines of a classical systems view, external observation that is obtained from the ortho-level, i.e., the level of direct observation, is inherently limited. The O(e) can only detect contradictions and distinguish differences in the o-properties of the system. External observation has yet to proceed (or infer) from the system's o-properties to its p-features. This intrinsic deficiency of classical cognition requires a necessary complementary view in which the O(e) is elevated to a meta-level of observation. This movement from an ortho- to a meta-viewpoint signifies a transition in handling a system on the basis of its *otherness* (ALLo) in order to perceive its *sameness* (AUTo). This marks the point where a hitherto systems observer turns into a *perceiver* of the system of concern.[18]

The Role of the Perceiver and the Participant

As demonstrated above, the role of the observer is not without serious shortcomings. A means of rescue from what could be called the "observer's trap"—an expression I recently coined in a critique of Niklas Luhmann's representation of "systems theory"[19]—is offered by turning to perception. The notion of systems perception is much richer than that of observation.[20] Above all, perception is not restricted to pre-formulated

---

17. Seemingly a way out of this dilemma, as proposed by Günther, consists in the turn of reflection proceeding from the "I" to the relation between the several levels of reflection; a trick which arguably restricts reflection to an operative procedure.

18. Locker, "Schöpfungs- und Evolutions-Problematik," 222.

19. Locker, "Angriff auf die ganzheitliche Welt-Auffassung."

20. The fact is emphasized in the assumption that perception exerts so-to-say the p-function of a system as against observation which is solely o-functional; it is thus

FIGURE 1.1: Allology—Autology Schema

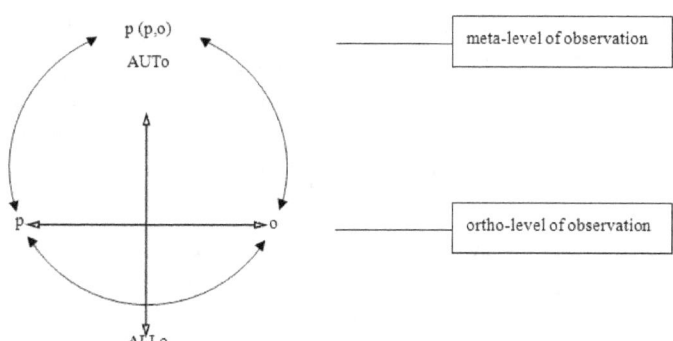

goals and thus synthetically overcomes the distinction between an endo- and exo-view of systems. In a manner of speaking, the systems perceiver is located on an elevated epistemic platform thereby embracing a view that is much wider than that of observation.[21]

Pondering deeper approaches to reality likewise involves the possibility of building models that allow for a certain immediate grasp of reality by means of widening and intensifying the mode of observation. Thus it does not seem completely exaggerated to think of experiencing reality in a three-, or even four-fold manner, such that a figurative sketch of systems views could not only relationally correspond to a circle,[22] but also to a cross-like figure uniting all possible systems views (O(e)+(i); perceiver; ALLo; AUTo). Circle and cross alike could be brought into mutual relation as it might equally be done with the notions of complementarity and correspondence, whereby both figures bear a complementary and corresponding relation to each other. The most significant finding in this comprehensive systems approach is that hereby the search for the physical origin of any system loses its meaning.[23]

the former which is able to encounter also the corresponding p-side of reality whereas observation is restricted to notice only o-results. This signifies another instance of the o-p correspondence in any system.

21. Locker, "Recent Approach to Transclassical Systems-Theory," 11–16.

22. Cf. Sloterdijk, *Sphären*, wherein the author replaced the widely used metaphor of the circle with that of the sphere which encompasses everything and simultaneously unites it with yet another thing.

23. Heidegger speaks here of "Gleichursprünglichkeit" (simultaneous originality), while Sloterdijk stresses the (at least) pair-wise togetherness of everything with another thing.

Perception establishes an intimate relationship with reality. Able to be adjusted and deepened, perception attains a high degree of harmony with reality. Perception does not "treat" objects from the outside, but rather occurs as passive—or better, sympathetic, even erotic[24]—encounter such that nothing of its own subject matter remains outside its own domain of cognition and recognition within the system. In the fullest sense, we can speak of depth perception rendering things "diaphanous" and revealing their divine side. Indeed, perception could be a key to the heavenly milieu (or even all-embracing ground) of everything. The English mystic William Blake (1757–1827) speaks enthusiastically of "doors of perceptions to be cleansed" which then reveal the Heaven on Earth. Goethe's concept of the "productive power of imagination" may describe something similar to this attitude; yet, it may be confluent with vision.

In the course of perception, the perceiver gradually changes into a participant in the system, who does not only passively perceive, but above all, sympathetically shares in the system's activities. We are thus justified to maintain an identity between perceiver and participant in the broadest sense, for the reason that neither drastically influence the occurrences they perceive, or in which they participate. On account of the truly damaging influences an observer in general inflicts upon the particular reality under observation, it would remain insufficient to term the perceiver simply a meta-observer. Likewise the participant must not simply be mistaken as active O(i). As has been expressed above, the assemblage of the two observers (as o-side), the perceiver and the participant (as p-side), forms yet another system, such that the ad supra stated theorem that any system requires a system for being grasped as such is corroborated.[25]

In sum, to approach a system only from the observer-perspective demonstrably proved to be a significant limitation in systems theory.

---

24. Marion, *The Erotic Phenomenon*. We encounter being, Marion says, when we first experience love: I am loved, therefore I am; and this love is the reason I care whether I exist or not. This philosophical base allows Marion to probe several manifestations of love and its variations, including carnal excitement, self-hate, lying and perversion, fidelity, the generation of children, and the love of God. Throughout, Marion stresses that all erotic phenomena, including sentimentality, pornography, and even boasting about one's sexual conquests, stem not from the ego as popularly understood but instead from love.

25. Locker, "The Present Status of General System Theory."

Recognizing, however, a system from a dynamic *perceiver-participant*-perspective—because of *in-*, and concomitantly *ex*-cluding the limited and limiting observer—will evidently suffice for approximating its true gestalt.

ORIGIN AND TRANSFORMATION OF SYSTEMS

The Problem of Origin

The aforementioned statement relative to the inaccessibility of the system's origin is almost certainly too emphatic, in as much as the problem of origin can equally be unfolded by stressing relationality. One senses intuitively that the problems outlined in the introduction of this paper are intrinsically related to each other. This impression is readily substantiated through phenomenology and daily experience. Albeit different and distinguished o-properties are needed to fully constitute a system, common sense tells that obviously a system cannot originate "piece-meal," but only at once and as a whole, for the simple reason that its recognizable gestalt is designated by its p-properties. Ascribing any quality to the whole system flows from the indissoluble unity of all of its properties, and would not be possible if they are fragmented.[26] Form this point of view, the notion of origin refers always to the totality of the system.

The event of origination, therefore, occurs outside of time, as it were, suddenly or so to speak in "in less than no time."[27] In apparent contrast to this comprehension, the act of observation is bound to last a certain amount of time and requires the lapse of time for its occurrence. This explains that despite the fact that observation remains a necessary yet supplementary condition for understanding the event of the origin, the method of systems observation on the whole remains incapable of touching the essence of the origin.

The notion of origin can be illustrated as a vertical arrow, exemplifying, for example, the sudden brain activity, called *Einfall* (the hoped for, yet unexpected and unforeseeable *invasion* of an idea), and has found multiple expressions in the various notions of the "origin of life," or better said, of the living organism, found in religion and philosophy.

---

26. A human can only be human, and not half-human, etc. Cf. the paradox concerning the inaccessibility of the moment when the tadpole becomes a frog: It never does, one moment it is a tadpole, the other it is a frog.

27. Locker, "Schöpfungs- und Evolutions-Problematik," 225.

The idea of origin interpreted as emergence from nothing (*creatio ex nihilo*) has often been criticized as a sort of "doubling," or "repeating" the concept of origin by way of transferring the notion of "idea" from the world of "ideas" into the world of "reality." Taking, for example, the machine-metaphor as somewhat important for elucidating the hereby suggested line of thinking,[28] we soon meet fundamental limits in this manner of understanding origin. Origin depicted as relationships between concepts does not provide for information on how the "idea" itself comes about. This "true" or "absolute" origin is perhaps bound to remain a mystery to human inquiry. Consequently, we must differentiate between a "hermeneutics" of origin—that can be represented by way of circular or cross-shaped schemas—and the dominion of "hermetics," revealing a reality we can only partially grasp by intuition, much akin to the notion of *Einfall* as described above.

Systems Change and Changing Systems

Change is somewhat different from origin since it proceeds from an already given and existing state. It can occur locally, affecting single properties of the system, and globally, changing the system as a whole. The notion of system as interplay-of-properties implies change. It is therefore inconceivable that during the "life"[29] of a system, i.e., the duration of its actions, its properties and characteristics remain unchanged. The changes of a system determine its course of history and can be broken down into different phases. One can rightly speak of an evolution of systems. Evolution in the true sense however, implies that whatever becomes apparent, emerges, or evolves, must already exist in a yet concealed form. It is these hidden intrinsic properties[30]—subsisting in the world of "ideas"—that, without being directly detectable, govern the

---

28. The machine-metaphor tries to demonstrate how the age-old philosophical problem of "universals" successfully applies here. Considering a machine, the "idea", i.e., the program for its construction, first materializes "ante rem" (viz. "in" the brain of the designer), then "in re" (actualized by the machine) and finally "post rem" (when an investigator, i.e., the O(e), explores the build-up of the machine). Speaking of the idea being "in" the brain, or of the program existing "in" the machine was only a *façon de parler*, in sooth does an idea (a p-property) only exist in the mind of man and is represented (or mapped) by material things (exhibiting for the observer only o-properties).

29. Because of the system's character as analog to the human person (and to organisms as well) this paper prefers the notion of "life" to existence or duration.

30. This figure is taken from Locker, "Evolutions," here figure 1.1.

evolutionary processes visible in "reality." These properties can only be inferred from the totality of the process. Living systems originate and evolve by conveying the "idea" of life and "change" with the purpose of animating and enlivening the material world.

Returning to the difference (and concurrent unity) existing within any system between the (global) totality and the (local) particulars of its elements, we can distinguish between changes of the whole—caused by changes of its parts, or a change involving the totality of the system at once—and changes limited to elements of the system. As constitutive part of the system, any given element (e) may undergo a change into another element (a), while the process of change may continue, either in a single direction (evolution) or in a reverse order (devolution).[31]

To further concretize these considerations, we have to ponder the following issues: (a) the phenomenology of change; (b) the reason why change and transformation occurs at all, including the mode of change, and (c) the goal (or purpose) of those occurrences.

A phenomenology of change is usually envisioned in terms of a one-directional transition of elements. This transition corresponds to the change of views in systems perception, and is comparable to the transformation of the unconcerned state of the "I" (the usual state of somber distanciation from reality) to an all-embracing, all-concerned, all-perceiving and enthusiastic state of "self." This advancement is somewhat similarity to the notion of *metamorphosis* and *transfiguration* described in philosophical and religious literature. All states of a system belong together. Their unity constitutes the gestalt of the system that—as we have seen above—can be understood as the entire system of reality or an excerpt (in the form of a partial system) of reality as a whole. Thus any transition or change of elements, or a sub-system, only reveals parts of the whole system. Other possibilities of change remain concealed and need yet to be actualized. In sum, only the totality of all possible systems changes or transitions will show the system in the state of wholeness.

From what has been said, one could get the impression that a change in the elements of a system simply pertains to its o-properties. This however would be a blatant misunderstanding, since those elements only touch on temporary states of the system. Perhaps it has not been

---

31. If the unity or harmony of elements in a system undergoes changes, a (sub)-system is constituted within the system. In turn, barriers between several elements of the system influence the dynamic processes that continuously occur in the system.

made clear enough that the transition of elements also refers to systems within systems (i.e., clusters of elements or parts of the system), and that both the local and the global aspects of transformation should be seen together. Then again, it should likewise be stressed that whatever any change achieves must by all means already potentially exist, such that the actual transformation is only an unfolding thereof.

Prior to asking why transformation occurs, considering the difference between transition and transformation will provide us with further insights in to what actually transpires in systems changes. *Transition* concerns stages or levels of a system. As they are targeted, and possibly also achieved, the subject (i.e., the system) undergoing the transition process remains fundamentally the same. In the event of *transformation* something more essential takes place. While reaming the same, the system (i.e., the subject) undergoes a profound change similar to an enduring state of self-transcendence. That something remains the same, although it is becoming the other of itself, is a paradox. This paradox can nevertheless be visualized (i.e., made visible by contemplative perception or perceptible intuition) as bodily and psychic transfiguration.[32] In this view the previously described transition from one element (e) to an*other* element (a) reaches its utmost completion.[33]

A possible way of speaking of the total transformation or *transfiguration* of the world is found in Romanticism. Novalis (Friedrich von Hardenberg (1772–1801)) and Friedrich von Schlegel (1772–1829) speak of the total "poeticization" of the world. However, the romantic notion of poeticization shows similarities to "postmodernism" and for that reason remains quite problematic. The main objection to be raised is that poeticization in the aforementioned sense exclusively relies on the power and potentialities of the human person without taking any recourse to the grace of God, which, in the view of this author, is an essential and thus indispensable part of the system of reality. It is this type of thinking that has led, for example, to the quest of gaining control over genetic dispositions, or the pursuit of cloning "better" human beings— not to speak of the promises of creating "Artificial Life" and "Artificial Intelligence" for the sake of improving humanities imperfect condition. Seen however in the light of grace, poeticization might even be realized

---

32. Cf. Locker "Leib—"höchste der Hieroglyphen," 194–218.

33. God's will to redeem his creation is the principal cause for the transfiguration. Cf. the treatise *Cur Deus homo* by Anselm.

as a first and necessary step against the hubris of the so called scientific achievement that gradually tends to displace the presence of God by mental power.

Metamorphosis and Transfiguration

The transfiguration of the system-subject can only be brought about through (a) a decision made deliberately and volitionally, i.e., a conscious and free choice, and (b) by means of a sacrifice in the deepest conceivable sense. The act of sacrifice is much more than an offering of something, or even the dedication of oneself to a certain task; it is more than self-constraint, or the offering of a gift. Sacrifice is self-renunciation, self-abandonment and self-surrender. The motive (or motivating force) for genuine sacrifice can only be love, of which St. Augustine says, grows the more it is donated. Again we do meet a paradox here: the more the system-subject is involved in this activity (composed of distinct single acts and also a lasting attitude), the more it gains for itself, such that self-loss is eventually tantamount to self-gain (the latter being the climax of the course of action in as much as this very same self is becoming the "true *Self*" within one-self; the divine kernel in every self; the Augustinian "intimior intimo meo," or the almost-identity of the human soul with God). Here a process takes place, whose essence we certainly miss in trying to describe it as mere identification of the individual system with the all-embracing universal system. Such identification is only secondary since each system is already—from a systems-theoretical point of view—representative of everything else, and thus all that there is. However the act of identification is necessary to bring about this change actively and consciously. Thus, the difference between a quasi-objective description of transfiguration and the true occurrence is marked by the involvement of a conscious decision and free will. Transformation and transfiguration does not occur automatically by means of obeying an unchanging natural law!

Regarding the purpose of transfiguration—i.e., item (c) in the above listing of the aspects of change, transition, and transfiguration—we can point to the inner tendency of reality to perfect itself (if we prefer not to speak of the will of God to redeem and sanctify his creation that Man can accept freely). In dealing with the modern tendency of reviving the ideals of Romanticism, the term "poetical metamorphosis," likewise coined by the key figures of this movement, can be understood in a positive way, largely because of its possibility to express the need for cooperation between God and Man. The final goal of all creation in general, and of human life in particular, can only be imagined as *apotheosis* (cf. John 10:35). Musing about the goals that are set by a system on account of its gestalt (as carried out by *teleogentic* system-theory[34]) implies the active pursuit of articulating and perhaps setting such goals within the system of humanity itself. This is primarily achieved by actively recalling that the possibility of creating such goals, and creation or co-creation within the realm of humanity, rests on the fact that all already has been created.[35]

THE LIMITS TO PERFECTION

A critical, yet sympathetic assessment of General Systems Theory (GST) unearthed a deeply embedded deficiency. Despite the fact that the mode of cognition employed by GST facilitates some excursions into the realm of the "productive power of imagination" by taking cognizance of p-factors—i.e., presuppositions hidden in the theories of "wholeness" and "gestalt"—by and large GST remains bound to formal logic and rationality. Strict rationality induces a deleterious inclination to assert exactitude, as it is most recognizable in GST's use of mathematical methods.[36] With full intent, GST observes criteria like strictness of proofs, consistency of deductive derivations, freedom from contradictions and the like, which all belong to the formal sciences, and thus, as argued already, cannot sensibly be applied to reality seen in its entirety.

Embedded in the whole of reality, i.e., in *real life*, we cannot avoid experiencing inconsistencies and paradoxes in the form of imperfections and perhaps inexplicable tragedies. These experiences elude the exact sciences and require interpretation in a broader philosophical and even theological sense; creation *hic et nunc* indwelt by humanity is undoubtedly an imperfect or "fallen" creation. Already early critics of systems-theories, like for example Johann Georg Hamann (1730–1788), caution against systems thinking by warning that any strict systematization could inflict harm on thinking like an instrument of torture. Systems

---

34. Coulter and Locker, "Recent Progress," 67–72.
35. Locker, "The Healing of Mankind's Predicaments," 131–52.
36. von Bertalanffy, *General Systems Theory*, 19.

would confine objects of thought to logical treatments and thus hinder truth from emerging.[37]

In this respect an important contribution was made by the Protestant theologian Friedrich Christoph Oetinger (1702–1782) who, among other things, not only rehabilitated the notion of the body, often despised in the history of the Christian faith, but also was extremely skeptical toward the concept of system that at his time was already commonly used for characterizing living things.[38] Oetinger thought that in this world no system—neither manmade nor created by God—could be perfect because the world is in disorder such that each system depicting it contains a *hiatus*, i.e., a gap or a fracture. Taking this fact into account will make us somewhat reserved regarding the capacities of GST and TCST, but by turning this observation into a positive claim we are invited to use the possibilities of these meta-sciences critically and positively.

---

37. Such comments have been made by J. G. Hamann (1730–1788) who criticized Kant's system by saying "System ist schon an sich ein Hindernis der Wahrheit," and by Friedrich von Schlegel (1772–1826) who addressed himself against Hegel stating "Es ist gleich verderblich, ein System zu haben wie keines zu haben." These expressions could possibly be understood as a hint for the need to ponder on the presuppositions for the system concept.

38. His well-known saying is: "Leiblichkeit ist der Anfang der Wege Gottes." For his system's view: R. Piepmeier, *Aporien des Lebensbegriffs seit Oetinger.*

2

Systems Theory and the Axis of Faith

Creation and Evolution—Resurrection and Life Everlasting

ALFRED LOCKER (1922–2005)

(translated and completed by Markus Locker)

ABSTRACT

The alleged incommensurability of religion and science is most evident in the search for meaning in relation to humanity's origin and destiny. General Systems Theory (GST) can bridge this disjointing by suggesting a multiperspectival approach on the basis of the notion of complementarity, i.e., unity found in diversity. The hereby implied view of reality stresses the epistemological link between the human person as complementary access system to all abstract systems and the necessary coping with paradoxes and forms of thinking and logic beyond classical means. A trans-classical version of systems theory (TCST) will argue that a system of First Things S(FT) and a system of Last Things S(LT) composed of allegedly unequal elements can indeed exist and coexist by recognizing that 1) everything contains the opposite of itself, 2) the existence of parts implies the existence of the whole, and 3) that self-creation is equal to the creation of something other than the self. On the basis of these assumptions the S(FT) and the S(LT) can be creatively advanced into a holistic image that is immanent and transcendent, concrete and universal, temporal and a-temporal at the same time, and hereby disclosing reality and truth simultaneously.

PROLOGUE

Throughout history humanity is engrossed with two questions: First, the origin of the human race, and second, humanity's destiny after life here on earth has come to an end. The second question evidently implies the conviction that the end of a person's natural life is not simply met by nothingness. Straightforwardly, these two questions explore the beginning and end (*archē*) of human life. While the natural sciences first and foremost provide answers to the first question—and faith, as it seems, plays hereby only a peripheral role—the second question without a doubt belongs to the realm of religion and theology. In view of the fundamentally different answers provided by science and faith to both questions, this paper assumes that a common basis for these allegedly incommensurate viewpoints can only be found if the general sciences are supplanted by the meta-science of General Systems Theory (GST), thereby providing for a possible connection between the empirical sciences and theological, respectively philosophical, reasoning. GST is believed to have the potential to build a bridge between the natural sciences and the humanities. By and large, the natural sciences deal with empirical facts, while the humanities or liberal arts are primarily concerned with questions of meaning and significance. Fact and meaning, however, belong together, and both have their respective place in GST.

On the basis of this hypothesis, GST can be appropriated for the study of topics involving questions pertaining to the notions of 1) creation, 2) an eternal world, and 3) a self-organizing or auto*poietic* cosmos. Together this cluster of questions can be viewed as a system of "first things," or S(FT). At the same time a coherent set of topics relating to the question regarding the *after*life can be outlined, comprising of concepts like: 1) immortality, 2) resurrection, and 3) life everlasting—taken together representing the so called system of "last things," or S(LT). At a closer look both thematic sets stretch far into the domain of the Christian faith where the postulates of the sciences and faith taken together pose contradictions that eventually build up to paradoxes.

It is believed that a study on the role of GST as possible link between the sciences and theology has not yet been undertaken, and already for that reason should attract interest. The use of GST hopes to effect a turn in public opinion that all too ready settles for accepting the popularly avowed incommensurability of science and faith. In order to convince the reader of the appropriateness of the arguments proposed here, GST has first to be adequately introduced as a meta-science that is capable of overcoming ostensible contradictions, intrinsic conflicts, and differences that surface within the classical conceptions of understanding "first" and "last" things.

The method used here remains strictly systems theoretical. The initial steps of advancing GST in this direction have already been undertaken.[1] Developing GST into a universal theory, however, will allow combining different—if not initially deemed mutually exclusive—ways of exploring reality, thereby including modes of cognition that normally do not belong to the realm of the sciences, like visionary sight and meditation. This somewhat modified form of GST has been termed Trans-Classical Systems Theory (TCST).[2] While it is strongly believed that the application of TCST proposed here does indeed call for a radical shift in systems epistemology, the contribution of TCST to questions concerning the "first" and "last" things intends to remain a modest contribution among many other equally qualified attempts in the direction of viable communications between science and faith.

THE ESSENCE OF TRANS-CLASSICAL SYSTEMS THEORY

It is necessary to begin this study with a definition—or at least a relatively accurate description—of General, and Trans-Classical Systems Theory. Both closely related theories endeavor to show reality (Wirklichkeit) in the most comprehensive manner possible. It must be said, however, that the insights of GST are exclusively derived from the premises of rational thinking.[3] Similar to the sciences, GST aims at the formalization of its results in abstract and general terms. What GST thus reaches is arguably not reality in its entirety, but—quite similar to the scientific fragmentation it seeks to overcome—only parts or fractions of it.[4] Notwithstanding that the birth of GST was stimulated by the pursuit of refuting a mecha-

---

1. A. Locker, "Recent Approach to Transclassical Systems-Theory," 11–16.
2. A. Locker, "The Present Status."
3. Bertalanffy, *General Systems Theory*, 47.
4. A fact that von Bertalanffy very clearly recognizes with regard to mathematical modeling: "There are highly elaborate and sophisticated mathematical models, but it remains dubious how they can be applied to the concrete case; there are fundamental problems for which no mathematical techniques are available," Bertalanffy, *General Systems Theory*, 20–23.

nistic understanding of the human person, because of its methodological foundation, it is believed that GST cannot fully claim independence from the classical sciences.

TCST tries to overcome this alleged shortcoming of GST, on the one hand by way of emphasizing the concrete, and on the other hand by trusting and following paths of knowing and knowledge that draw from imagination and the power of vision—as for example found in the works of Johann Wolfgang von Goethe and Robert Musil. For this reason it is believed that TCST is able to surpass the intrinsic limits of one-dimensional logic, thereby successfully dealing with paradoxes and the notion of ambivalence. TCST explores areas of reality that in conventional or mathematical systems theory are deemed non-, or un-scientific, and seeks the active dialogue with any and all forms and expressions of reality based on human experience. Reality, it is believed, never shows itself in only one, single, and consistent mode.

TCST stresses first of all the universal analogy of system and reality. This analogy is primarily manifested in the fact that any system appears as an entity that in some parts shows characteristics similar to its designer or the person who conceptualized it, while in other parts appears to be entirely different, obtaining a life of its own and a certain autonomy. According to Ludwig von Bertalanffy (1901–1972), the acknowledged founder of GST, the general paradigm of system is the biological organism seen and understood as a whole or *ganzheit*. TCST surmises that this whole is not simply found in the system's organization (i.e., the sum total of its elements and processes), but embedded in its *gestalt* or Idea. The gestalt of a system encompasses all its characteristics by way of analogy to the reality from which it is distinguished. The notion of systems gestalt denotes all possibilities of reality to be represented in systemic-holistic manifestations. On the basis of this conception, systems can initially be understood as clusters of special properties (like, for example, chemical substances and their processual coherence) that sustain themselves in a particular configuration against interfering influences originating from an environment. The gestalt of a system accounts for its autonomy that occasionally can be very pronounced. Some artificial systems appear to have a life of their own. Yet TCST assumes that any systems designer must eventually accept that no designer creates life and only actualizes the already-given possibility for it. The designer—like his own design—is merely a part of reality, likewise taking on the form of a system.

Over and above this, the concept of systems analogy recognizes a quasi, or virtual identity (autology, or sameness in essence) of designer and system. This systemic equality establishes an implicit and/or explicit dialogue between the system and its designer. Systems do not only interact with their environment, but by necessity with their designer. This important realization requires the critical evaluation of the relationship of system and human beings in general. What on account of observing and studying systems is deemed real or reality, here and now (*momentane Wirklichkeit*), must not be confused with what is true or actual reality (*eigentliche Wirklichkeit*).[5] TCST believes that true reality is perceived when the experience of the designer is brought into his design. For example, the Christian faith surmises that our momentarily experienced reality is but a temporary and fallen reality (a systemic extract of true reality), distinct from, and embedded in eternal perfection surrounding us and this world.

On the basis of the aforesaid, it becomes clear that a person who distinguishes and scientifically describes a system is equally the designer of its underlying theory—the systems theory. In the case of biological or social systems, the designer obviously uses this systems theory (for example the theory of biological life, or society) to further understand and describe the system of concern. Any single perspective or point of view of the system will, however, remain an insufficient vantage point for understanding its true gestalt. TCST proposes that in order to perceive a system's entirety, the designer must assume a variety of different positions relevant to the system. Initially, he will be its *observer*, and after obtaining sufficient understanding, he can become its operator. Realizing the limitations of the observer position (outlined below in more detail), the designer, on the basis of his essential analogy to the system, may begin to *perceive* it. Perception provides the ground for intuitively entering into the system, where the designer finally comes to realize himself as *participant* in the system's activities. Taken together, these three system views overcome the observer's blind spot[6] and provide the ground for *beholding* the system. From the point of view of a beholder, a system that in fact can never be observed objectively can yet be beheld in its entirety. The beholder is permitted a holistic vision of the system. In sum, TCST differentiates four basic system viewpoints, all essentially necessary to

---

5. This differentiation alludes to the ontological difference.
6. A. Locker, "Angriff auf die ganzheitliche Welt-Auffassung."

grasp the gestalt of a system: 1) observer, 2) perceiver, 3) participant, and 4) beholder. These four positions are joined together in one person, and account for what must be understood as complementary human access system (CS) to reality.

All four systems viewpoints are essential in ensuring a glimpse of the system's whole or gestalt. Yet, each perspective displays unique insights that cannot be placed in concert on one and the same level of logic and understanding. For example, the concept of systems observer denotes control over the system on the basis of mastering its theory. The limitation of this view is unmistakably clear. To the observer the system only exudes the very segment of its reality that he has already imposed onto it by means of the observation.

It is well known that an observer can obtain a position outside and/or inside the system.[7] Yet even an inside, or endo-view of the system (sometimes associated with second-order cybernetics[8]) is bound to the limitations of observability, and consequently allows the system only to show what conforms to the observer's own interests and prejudices. Accordingly, any "other" reality of the system will be ignored, and more often than not, is categorically denied. TCST believes that misrepresenting any system in this way and consequently subjecting it to one's own interests actively prevents the actual gestalt of the system to be revealed, or to reveal itself. This fact is indicative of the so called "observation dilemma."

Perceiving of a system is very different from common observation. In a manner of speaking the perceiver moves to the background and allows the system to reveal itself. Systems perception in this way overcomes the apparent difference of the inside and the outside of a system. Perception occurs within the *difference* and encompasses both sides of the system, hereby giving preference to the endo-view. Systems perception leaves the realm of analytical thinking and moves to a *synthetic* view of the system.

In general, the same can be asserted for the so-called participant view. Participating in a system establishes accord and harmony between observer and system. The participant enters into a close relationship with the system and is acknowledged as an integral and constitutive part

---

7. Rössler, "Endophysics," 154–62; and M. Locker, *Systems-Theoretical Considerations*.

8. Foerster, *Observing Systems*.

of it. This view likewise transcends the mode of observation of designer and operator. While the latter is bound to rationality, perceiving and participating in a system includes *intuition* and *imagination*.

In the search for the gestalt of a system, TCST realizes that over and above all systems viewpoints, one can, and must conceive of an absolute beholder (Betrachter) of a system. The notion of an absolute "observer" is far more than simply the sum total of viewpoints, but found in the necessary realization that conscious systems differentiation cannot be rooted in an infinite regress of "systems observing systems."[9] The beholder viewpoint is akin to a systems view inside the ultimate environment of all human viewpoints and must be envisioned as obtaining a measure of indifference, transcendence, and completeness, i.e., a vision in which all hitherto existing systems views are grounded.

It is not difficult to realize that the relationship between these system viewpoints constitutes itself a genuine system. This human access system to the system of concern can be regarded as accompanying or complementary system (CS). Despite clear differences between the system proper (S) and its complementary system (CS)—whose basic elements are the aforementioned 1) observer, 2) perceiver, 3) participant, and 4) beholder properties—both systems are part of the one true reality (egentliche Wirklichkeit) that is only fully represented when the properties (Eigenschaften) of both systems are viewed together. If a single constitutive element of the CS enters into a relationship with the system proper in isolation, only a limited view is achieved. Any partial systems cognition is in need of being complemented by yet another, if not all possible systems viewpoints.

Since the complementary system—the human person—can observe itself (i.e., can speak of him/her-"Self"), so too the system proper has to be granted a "self."[10] The self-sustaining elements of both systems com-

---

[9]. The lynchpin of Luhmann's systems thinking is the doctrine of the blind spot: "Ein Beobachter kann nicht sehen, was er nicht sehen kann. Er kann auch nicht sehen, daß er nicht sehen kann, was er nicht sehen kann. Aber es gibt eine Korrekturmöglichkeit: die Beobachtung des Beobachters. Zwar ist auch der Beobachter zweiter Ordnung an einen blinden Fleck gebunden, sonst könnte er nicht beobachten. Der blinde Fleck ist sozusagen sein Apriori. Wenn er aber einen anderen Beobachter beobachtet, kann er dessen blinden Fleck, dessen Apriori, dessen 'latente Strukturen' beobachten. Und indem er das tut und damit seinerseits operierend die Welt durchpflügt, ist auch er der Beobachtung des Beobachtens ausgesetzt." Fuchs, Peter und Niklas Luhmann, *Reden und Schweigen*, 10–11.

[10]. While the human "Self" exists in independence from its environment, the sys-

plement one another, but at the same time appear as contradictions. This palpable inconsistency between CS and S is not accidental or contingent, but a genuine aspect of both systems and the hereby necessary notion of systems complementarity. This will be clearly seen in the attempt to understand the complementary relationship between the S(FT), the CS, and the S(LT). In view of their complementary the elements that are found in the CS and the elements of both systems (like creation, timelessness, and evolution within the S(FT), and immortality, resurrection, and eternal life in the S(LT)) constitute genuine sub-systems, each of which can be considered independently.

COPING WITH CONTRADICTIONS IN THE *SYSTEM OF FIRST THINGS* S(FT)

The system of the beginning of all existing things, commonly summed up in the notion of creation, and in the alternative concepts in the S(FT), is verily compatible with the notion of God as the "designer" of this system.[11] Despite apparent and significant contradictions, the properties constituting the S(FT), i.e., (a) "an eternal world," (b) "the (*ex nihilo*) creation of the world," and (c) the "self-creation (auto-poiesis) and evolution of the world," can be envisaged as complementary elements if certain assumptions (that perhaps initially appear as paradox or absurd) are made. These would be:

1. Any object (Seiendes, or entity) is not only that which it appears to be at any given moment, but likewise contains the quality to be transformed (in the sense of emergence or metamorphosis) into the other (side) of itself. This consideration implies the understanding that a concrete and single entity to a certain degree represents—or even is tantamount to—all of existence.

2. This understanding further implies that the appearance of a singular object already signifies the emergence of its totality. Alexander von Humboldt captured this truth by maintaining that together with the first word all of language came into existence, and Johann Wolfgang von Goethe conceived of the

---

tem's "self" subsist alone in relationship to its human complementary system. This accounts for the use of upper and lower case.

11. Not to be mistaken with the notion of Intelligent Design.

notion of an Urpflanze, or archetype, of all possible plants in the form of a combinatorial entity representing all the varieties of existing and imaginary plants.[12]

3. A third necessary assumption, in view of the fact that autogenesis is the necessary precondition of all allogenesis, is that self-creation (that in the case of God corresponds to God's eternal being) is equal to (or, the same paradigm as) the creation of something other than the self.

In accepting these paradoxes light can be shed equally on the S(LT) by way of its conceptual similarity to the S(FT). In the S(LT) the notion of self-transformation as a consequence of the unlimited interconnectedness of its elements reaches its definitive perfection and consummation. The foregoing conceptual assumptions are likewise the precondition for resolving the contradictions found in both systems respectively. In the cold light of day the individual elements of the S(FT) (and as seen later likewise the elements of S(LT)) contain ambiguities and contradictions. This can be demonstrated through the notion of *creatio ex nihilo*. Some two hundred years ago the Baltic German biologist Karl Ernst von Baer (1792–1876) maintained that the very concept of creation out of nothing already presupposes the existence of an eternally existing Idea of creation (akin to the Holy Spirit, i.e., the vital or animating principle of God). Equally, the notion of an eternal world includes the difficulty that only on account of continuing creation (respectively the creative and sustaining "thinking" of the world by God) can such a world be held in existence. In this vein, Johannes Kepler maintained that one cannot think that geometry is not co-eternal to God.

The same is true for the assumed self-creation of the cosmos, as this concept likewise presupposes the notion of a personal *Self*. The postulate of a systems' self on account of its auto-poiesis remains, however,

---

12. Goethe's superior imagination allowed him to actually see the *Urpflanze*, or proto-plant, containing the essence and *gestalt* of all plants. In a letter to Herder he explains: "The primordial plant would be the most wonderful creation of the world, for which nature itself should envy me. With this model and the key that it contains, one could invent an infinite number of plants, ones that despite their imaginary existence could possibly be real, thus which are not solely literary and painterly shadows and illusions, but which possess an inner truth and necessity. This same principle would be applicable to every other aspect of life as well." J. W. von Goethe, *Italienische Reise* in a Letter to Herder on May 17, 1787.

forgetful of this truth. In reality, anything new in nature can only be perceived as such in relation to an already existing concept of newness or an equivalent final goal, i.e., in the case of the S(FT) the cosmos itself.

The existence of a personal *Self* as noun is likewise the condition sine qua non for the pronoun *self*. It is the very self-designation of God (Exod 3:14) as *Self* that allows for the creation of the self. This circumstance serves as an example for the hereby ensuing and likewise paradoxical assumption that *nothing* takes place here and now that did not already take place in eternity, whereby at this juncture the objective event (of the *here*, or *immanence*) must be thought in unity with its presupposition (*transcendence*).

How can these contradictions and paradoxes within the S(FT) (and thus also the S(LT)) be resolved? First, it must be remembered that the elements of the S(FT) obtain equal value among each other, and that the system itself can be seen as the expression of a transcendental idea (in the sense of a Platonic *Idea* interpreted within the framework of the Christian faith) that in itself is above and beyond time. This idea, for example, is the condition for the temporal and tangible event of evolution.[13] In order to substantiate this concept we have to turn to a deeper consideration of the notion of system and the specific behavior of systems properties.

System and Concept

Up to this point we reflected upon the concept of system solely from the point of view of a cluster of elements. However, with regard to the earlier-established analogy to the human person, systems in a manner of speaking have an infrastructure. This point can be illustrated from the perspective of the beholder. At first glance it appears that all systems are founded on two different realities; on the one hand systems are distinguished on grounds of their objective aspects (termed G-properties, referring to the German word *Gegenstand*, i.e., object), while on the other hand the concept of system rests on and flows from its pre-existent theory or *presuppositions* (i.e., the so-called V-properties, derived from the German term *Voraussetzungen*). Initially, G and V properties (like for example G (eternity and temporality) and V (sameness and change)) seem to oppose each another. At a closer look however, and in widen-

---

13. Van der Meer, "Alfred Locker's Critique of Evolutionary Thought."

ing this understanding through the notion of systems perception, these oppositions can be joined, and gradually come to constitute a *unity*. G and V perspectives can likewise be characterized as different levels of systems accession. Resolving the tension between oppositions, unity is found in the notion of *complementarity*. Complementarity is the key feature of the self-activity (Eigenaktivität) of any and all systems. Systems complementarity expresses the unity behind the so-called ontological difference, i.e., between Being (Sein) and beings (Seiendem) that arguably permeates all reality.

Complementarity

The concept of systems complementarity can be further developed on the basis of a superordinate *tertium*. This can be schematically illustrated in terms of the difference and unity of V and G. The system itself is hereby the unifying third principle. In this schemata complementarity can be represented as unity of V+G and formalized as meta, or complementary presupposition: $V_c(V+G)$. While contradictions and oppositions within the system have been explained in terms of the inherent otherness (*allology*) of a system to its environment and the complementary system CS, the accordance of its properties is known as systems *autology*. The concept of systems *autology* closely relates to its gestalt or ganzheit.

The most common V system characteristics can be summed up in the acronym AEIOU:[14] A, for *autonomy*; E, for *existence*; I, for *individuality* and *identity*; O, for *order*, and U, for *unity*. Taken together, these properties represent the quasi-essence or inner nature of the system. Examples for G-properties are all of the system's tangible entities, such as its elements, structure, function, hierarchic, or flat design and the system's physical activity.

The infrastructure of the S(FT) is dominated by the contradictions posed by its elements. These contradictions take on sub-systemic forms and render the system into an *allo*-system, i.e., a system that in essence abounds with contradictions or conflicts. Then again, the shift introduced earlier in observer positions allows for transforming these disagreements within the system. On the basis of systems *autology*,

---

14. The Austrian Emperor Frederick III (1415–1493) habitually signed buildings and objects with this acronym. Cf. Kohn, "AEIOU," 513–27.

contradictions in the S(FT) take on the form of complementary properties and now account for its unity.

Systems complementarity begs the question of which properties in the S(FT) must be understood as V-, and which as G-characteristics? Since all its elements (with respect to their position towards its designer, i.e., is God) obtain equal value, one must assume that each element can obtain the role of either a V-, or a G-property. This insight does not lead to any further insights in interpreting the system in question, yet reminds us to consider the relationship of the complementary system (CS) to the system proper, i.e., the S(FT). The two systems positions that do not disturb the *autology* of the S(FT) are "participant" and "beholder." Combined, these systems views allow for a deepened access of the system via observation, creative imagination, meditation, and contemplation in the sense of mystic ecstasy.[15] In this way the harmony of the system becomes visible through archetypes of its origin (*Urbilder*) in conjunction with its temporal evolution, although only for moments of utmost ecstasy and the bliss being granted the vision of God.

Evolution

For that reason, evolution—deemed by the sciences as actual creation—is realized to occur merely in terms of the G-properties of the system in opposition to corresponding V-properties, or as the role performed by the elements of the S(FT). Taking into account that serious scientists have interpreted evolution in terms of the Fall of humanity,[16] evolution must be understood as a complementary feature of the S(FT). Originally, humanity does not find itself in the visible world. Human exclusion from the world of God as the proper designation of the primordial system and in which humanity remains a constituent participant, is poetically viewed as the origin of sin. On account of sin, humanity loses its original *gestalt*, i.e., being fully indwelled by God's glory, and has to retrieve its true nature by passing through many millions of years of evolution from lower organisms, until its re-covered and re-constituted Self once again appears, first in early humans, then culminating in the events of the Incarnation, suffering, death, and resurrection of the perfect and first

15. A. Locker, "Der Mensch: Nicht unbeteiligter Zuschauer," 34–42.

16. Evolution seen as result of the disobedience of the first couple succumbing to the temptation of Satan, the Evil one, somehow causing creation to fall from its primordial status to which it slowly moves again through evolution.

human, Jesus Christ, in whom the once torn bond between humanity and God is finally restored. In terms of the goal of evolution, humanity in the material world is initially merely present in the form of creatures. Envisioned in the Cabbala, Adam Kadmon,[17] the perfect Man before the Fall, and in terms of TCST, the beholder and at the same time participant in this occurrence, perceives this passing and evolving of the fallen Man, or Adam Belial, i.e., the wicked Man, through the animal kingdom that is understood as containing rudimentary humanity, and yearns for the reappearance of his absolute nature. From the point of view of TCST, humanity already existed during early evolution in a twofold manifestation: 1) in an actual but imperfect gestalt within the realm of the animal world, and 2) potentially as perfect humanity as its goal. This circumstance remains likewise in harmony with the presence of a supra-temporally/a-temporally existing S(FT), all the same not resolving, but genuinely preserving its contradictions.

Resolving Classical Paradoxes

Drawing from the creative potential of systems contradictions in the attempt of harmonizing them can in fact intensify the apparent paradoxes within the S(FT). This, however, is neither necessary nor desirable. Yet there exists a manner of deepening the complementary and relational basis of contradictions. Inconsistencies that are perceived as perturbing can be differentiated by means of reference to the classical segment of the system. In this way the disturbing effect of this systems property is disabled and transferred to a trans-classical part of the system that functions as synthesis of the paradoxical and non-paradoxical realms of the system. This transfer results in the alleviation or resolution of the paradox, since the latter is now profoundly integrated with the system and becomes its incentive. Hereby, the system itself achieves a higher form of identity (Identitätsspannung) through the effort of overcoming the paradox that comes along with the system's differentiation.

A similar effect can be assumed with regard to the conceptual presence of a mystically disposed participant, whose presence some-

---

17. "According to the Kabbala the first man [is] an emanation of absolute perfection. He is symbolized by the major axis of ten concentric circles, the Sephiroth or the ten circles of creation. Thus, Adam Kadmon as primeval man symbolizes the universe. He is androgynous, and is seen in ancient Jewish mysticism as partaking in, or blending with God." Lurker, "Adam Kadmon," 6.

how causes the inconsistencies in the S(FT) that originated in God to be eliminated. Contradictions that are mentally-meditatively resolved by the intellect of the participant are overcome in a way that allows the system to acquire an *auto*-Gestalt. At this point we can likewise consider how inconsistencies can be alleviated in the System of Last Things (SLT).

RESOLVING CONTRADICTIONS AND CONFLICTS IN THE *SYSTEM OF LAST THINGS* S(LT)

The elements forming the system of last things, viz. (1) the immortal soul, and (2) resurrection and life everlasting, are not as sharply opposed to one another as the elements of the S(FT). Yet each of these components carries along significant contradictions. The immortality of the soul, for example, cannot be imagined without corporeal manifestation that above all remains the guarantor that the identity of the departed person is sustained until one's final resurrection. The notion of the resurrection itself appears similarly ambivalent. Several hundred people witnessed the risen Christ (cf. 1 Cor 15:6). The final resurrection however, taking place at the time of God's final judgment, will have no witnesses of this world because by then it will have come to an end. And how will those raised differ from those living in all eternity like the angels? An additional enigma is found in the apostle Paul's declaration that in the final coming of Christ, those who are still alive will be taken up (*harpzō*) immediately, and in this way are granted a favor even greater than that of Jesus and Mary, the Mother of God (cf. 1 Thess 4:17). Along these lines, the resurrection could be understood as the final return to the original being; as Adam Kadmon, regaining, and even superseding, this form or gestalt of existence.

These considerations lead us to the concept of everlasting life (EL) and the difficulty in envisioning such a state without significant inconsistencies. Since the resurrected existence implies a definitive form, it ought to include the capacity to display all manifestations of our being in this life at once, or, in the words of Novalis, "the universal capacity of all natural things to be wine and bread in eternal life."[18] One can perhaps as-

---

18. "Das Christentum ist dreifacher Gestalt. Eine ist das Zeugungselement der Religion, als Freude an aller Religion. Eine das Mittlertum überhaupt, als Glaube an die Allfähigkeit alles Irdischen, Wein und Brot des ewigen Lebens zu sein. Eine der Glaube an Christus, seine Mutter und die Heiligen. Wählt welche ihr wollt, wählt alle drei, es ist

sume that such existence allows us to appear in all our previous appearances, and according to our desire even in the form of wind, cloud, or any creature. EL may be characterized by this playful metamorphosis and on the basis of retaining the capacity to assume any arbitrary form remains to be sharply distinguished from the so-called metempsychosis or transmigration of souls understood as the inevitable consequence of personal fault. In fact, this transformation is an expression of redemption.

Yet another ambiguity inherent to the notion of EL must be acknowledged. If eternity has no beginning, EL likewise must belong to the human person from the beginning, however with the imperfection of the possibility of suffering its loss temporarily, resulting in humanity's postlapsarian existence. Such being the case, eternal life bestowed on us after the resurrection signifies a return to the origin, yet surpassing it.

At this point, we increasingly become aware of the deep connection between the S(FT) and the S(LT). In fact, there exists an analogy, but likewise a distinction, between the S(FT) and the S(LT). While the complement to the S(FT) is evolution, there seems to be no analogous complementary systemic property to the S(LT). We recognize this difference considering the equal positioning of the elements within the S(FT). There seemingly exists no equal status of properties in the S(LT). The defining property of the S(LT) is the element of EL that perhaps corresponds to the eternal world in the S(FT).

Resurrection, like creation, is a momentary event pointing to yet another correspondence between these two systems. Self-creation remains without analog in the S(LT), unless the essential element in EL, i.e., transformation, is understood as self-effected. Furthermore, EL has no temporal dimension like evolution in comparison to the S(FT), but takes place a-temporally, temporally, and supra-temporally, permeating all events in EL and bestowing a paradox onto it. This paradox is, however, genuinely overcome in the participant of EL, finding himself in a synthetic time-system and not perceiving this paradox, which is only visible from the outside, or the theoretical observer perspective. At this point it must be remembered that already this "fallen" existence contains the very same time paradox experience, aptly contained in the expression that "any moment at the same time represents eternity." This

---

gleichviel, ihr werdet damit Christen und Mitglieder einer einzigen, ewigen, unaussprechlich glücklichen Gemeinde." Novalis, *Die Christenheit oder Europa*, par. 24, 1799.

insight signifies a deep level of awareness of any and all interconnections of reality, and especially of all "first" and "last" things.

As regards to the S(LT), and because of the unequal value of its elements, one must assume that EL as culmination-of-human-existence obtains a special role, viz. the transformation of the S(LT) into the system of EL. The paradoxical return of humanity to its transcendent origin flows from the correlate of the G-side of the S(FT). If EL does not imply forgetting past experiences, and can be imagined as being saturated with memories of actual events, heaven and the *new* earth may be envisioned as *play*ground for experiencing love and transformation, and for newly living the life of the past, herewith solving all hitherto insolvable conflicts and surpassing all previously unsurpassable problems. EL would contain a transformed memory of humanity's earthly existence. In this view, the S(LT) too can contain the equivalent to the complement of the S(FT), i.e., namely the events of the past that are playfully relived hereby alleviating, or *re*solving what previously in our earthly existence appeared to be irreconcilable contradictions.[19]

A playfully relived life introduces a dimension of time into the three-time-system of eternity that can be termed *memory*-time. This time is complementary to EL, but in difference to (temporal) evolution belongs to the supra-temporal S(FT), and to the *eigen*-time bound to it. On the basis of this point of view, we recognize as preliminary result of our study that the relationship of the S(FT) to the S(LT) is *symmetrical*—as the S(FT) is a condition for the emergence of the S(LT)—and at the same time *asymmetrical*. S(FT) and S(LT) are at once *allological* and *autological*, since there exists a correspondence between both, i.e., the beginning and the end, which in both systems are those reference points that constitute reality.

FURTHER CONTRIBUTION OF TCST TO THE CONUNDRUM

The assumption that each designed system is, to a certain sense, an explication of what is known to its designer at the same time containing its entirety and *gestalt* is the precondition for the theoretical construction of the S(FT) and the S(LT). Since, however the interaction and

---

19. "In a vision St. Perpetua saw her brother Dinocrates, who had died from a disfiguring disease and unbaptized at the early age of seven, in a place of darkness and distress. She prayed for him and later had a vision of him happy and healthy, his disfigurement only a scar." Kirsch, *Sts. Felicitas and Perpetua*, 23.

dialogue between the system proper (S) and is complementary system (CS) likewise belongs to this process, one has to admit the necessary existence of what can be called the *instance of faith* for the conception of S(FT) and S(LT).

From the standpoint of the CS, the S of concern appears as "the *other* of itself." This implies that this *otherness* is revealed through the CS on the basis of its peculiar inconstancy. This marks the outer limits of its capacity in understanding the system of concern. The CS becomes fully aware (*intro*-cendence, as complementary activity to *trans*cendence) of S by recognizing the correspondence of its origin with the origin of the S. Both systems, in this way, mirror each other. In turn the CS's capacity for transcendence must be seen as the necessary condition for its existence, continuation, and capacity for change. Since all systems are *analog* and *analogs* of the CS, the S(FT) and the S(LT) are not only analogous systems but in similarity bound to the human person, who is not simply fated to remain a passive participant in the systems of S(FT) and S(LT) but revealed as systems designer bestowed with the power of visionary imagination.

At this point it must be emphasized that in the case of creatively designed systems, contradictions are only found here and now, but not, for example, within EL where these contradictions are actually lived and therefore not constant and binding. In looking back at our past life, the presumed eternal "love inter-play" consisting of constant metamorphosis can intensify into an intellectual interplay, conceiving time and again of new paradoxes and musings.

In general, contradictions and tensions surface when the system contains elements that are neither fully separated, nor truly unified, like, for example, the aforementioned AEIOU properties. These elements allow for the *ontologization* of the system but at the same time bring about a certain imperfection and the condition of tentativeness. This insight brings to mind that all statements about systems aiming at truth-and-reality must be constantly held in suspension, and not become definitions.

This, however, only applies for reality systems prompted by the CS, and not for systems surrounding reality, like the S(FT) and S(LT). Contradictions and paradoxes belonging to this, i.e., our reality, exist in it as disturbance (irritation) and therefore act as necessary inner stimulator of any system. The necessary existence of such cannot be assumed for the S(FT) and the S(LT). While these systems are in reality free of contradictions, they appear as containing them because they are conceived in the human mind and from a worldly point of view. S(FT) and S(LT) are able to remedy contradictions and paradoxes by either expanding the context of the paradox, or by deepening its basis, which corresponds to the aforementioned *intro*-cendence of the system to the "Self."

Through the resolution of paradoxes, the meditative participant represents God, whom we can think of as the origin (and end) of all contradictions found in reality and what lies beyond it. "Releasing" the paradoxes of EL, through, for example, the vision of the prophet Isaiah (Isa 11:6; 65:25) where wolf and lamb, leopard and kid, and lion and calf peacefully rest together, can be developed into a general paradigm for *supplanting* paradoxes. This biblical image represents the mutual transformation of one into the other (metamorphosis) with the effect that no creature has to starve to death; which would be possible in this world only if a lion were forced to graze. The lion—envisioned to be transformed into a lamb—, however, would be able to feed on grass and to retain its strength and vitality. In the S(LT), changing and metamorphosing into EL points to an existence in ingenuous and playful interchange.

Admittedly many objections can be raised as regards to the foregoing illustration, the more so as it remains questionable if the elements that are united in the S(FT) and S(LT) actually become properties thereof. In general, one has to make the needed distinction between *formal*, i.e., theoretical, and *real* systems. The former are complete in their formal structure and therefore free of internal contradictions, while the latter always suffer from inconsistencies. Yet real systems are characterized by autonomy and their *near complete* identity with the CS. Furthermore, one must remember that real systems not only seek to demarcate themselves from their environment, to which also the designer belongs, but as a part of reality also show the tendency to merge with it.

The S(FT) and the S(LT) are certainly not created by their complementary properties. Both reveal God as their designer. In this way, both systems are super-systems (über-systems). Even after the reality here and now has passed and is surpassed by EL, one can remain to see the S(FT) and S(LT) as building blocks or mainstays of a final super-system in which both systems are fused. While this entirely new system, or *New World*, is seemingly out of reach of our consideration, it can nevertheless be illumined by applying TCST to some of its properties. Humanity, for example, which is a participant in this system, will find itself throughout the entire system: (a) within an a-temporal pre-, or ur-existence, (b) within a temporal intermediate-existence (in the form of an imperceptible passing through a pre-human animal-existence up to his return to his human *form*), (c) in a post-temporal existence (as "soul" in the afterlife between individual death and final resurrection) and finally, (d) in a definitive pan-temporal (a-temporal/supra-temporal/temporal) existence in EL. This realization urges us to redefine our roles here and now in recognition of our true origin and end by fully cooperating and participating in the destiny of this world.[20]

Yet another question that has to be raised is the status of *us* who undertook this study in relation of our foregoing considerations. As far as it can be recognized, all propositions discussed above conform to the magisterium and must not be misunderstood as a new type of gnosis. Thus, it is hoped that at long last this short presentation will generate some interest among open-minded theologians.

Finally, in returning to the introduction, it must be pointed out that the realties and processes herein outlined and systems theoretically examined would lack significance and meaning for us here and now if one would fail to notice the central significance of the S(FT) and S(LT) for the relationship of God and humanity in general.

EPICRISIS

Among many other things, what perhaps remains wanting in this study is the emphasis on a transcendent God, who though designing, i.e., creating, the S(FT/LT) remains unambiguously transcendent. In view of reality here and now, the system proper (S) and the CS are on an equal level. Open to artistic inspiration and imagination, human beings can and do comprehend the reality of creation flowing from the S(FT) and the vision of EL generated within the S(LT). Any theory-building however must remain open to ongoing changes and modifications. Actively touching what lies beyond a first and superficial glimpse of reality continues to deepen the conviction that humanity genuinely partakes in all of reality.

Second, in introducing the notion of metamorphosis, emphasis must be given to the joy of salvation and the idea of the possibility of

---

20. A. Locker, "Healing of Mankind's Predicaments through Sufferings," 131–52.

play and *dance* before God. Only in this way the quest for knowledge can overcome despair, hereby truly revealing systems theory as constructive and helpful epistemic tool for humanity.

Finally, the exchange of the S(FT) and the S(LT) with their environment brought about by the systems theoretical enquiry of the human CS increases the inner dynamics of both systems that hereby expand and develop. Indeed, the entire super-system (S(FT)/CS /SP/S(LS)/EL, i.e., the *New Reality* as transformed unity of previous theoretical and real systems) possesses vitality and thus can be conceptually advanced indefinitely. This system emerges as a vision held together by systems transitions (S(FT)-*incarnation*-CS/S-*resurrection*-S(LT).

It is our Christian belief that in and through baptism we are empowered to transcend the systems of *hic et nunc*, temporarily through the bliss we receive through images of reality found in religion, art, and philosophy, and finally by being granted a beatific vision where in the words of the apostle Paul we will see God face to face (1 Cor 13:12).

FIGURE 2.1: The Super-Systems of *First & Last Things*

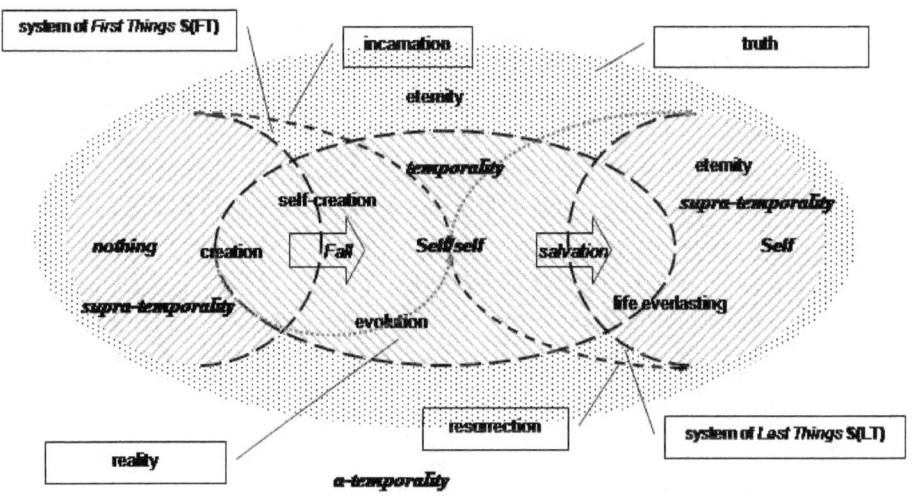

3

Edenology—A Science of Paradise?

OTTO E. RÖSSLER

ABSTRACT

Does an Edenology—a scientific investigation into the question of whether or not we are living in paradise—make sense? The idea goes back to Markus Locker who wondered whether or not the most original and least easy-to-understand Jesuan parables may deserve to be taken literally as truthful statements in the sense of modern science. Three scientific topics will be re-addressed with an eye to this question: The status of the problem of the Now, the theory of mothering, and Everett's quantum worlds. The two unfamiliar technical terms that are identified are "assignment conditions" and "benevolence theory." The former notion needs to be added to the 320-year-old technical notions of "initial conditions" and "laws." The latter revives the forgotten technical term of the "sun of the good." The existence of assignment as a fact of nature calls for a better understanding of its nature, acceptability, modifiability and, most of all, intention.

INTRODUCTION

MARKUS LOCKER HAD THE almost paradoxical idea of taking a religious statement literally as a scientific statement: "The kingdom of heaven is amongst us." It is related to another traditional saying: "The smile of Buddha is all around us." Is the spirit of this kindred claim just a psychological solace, an effective "opiate for the masses," or is it a provable fact in the sense of modern science?

"YOU are in the scent of the flower" is a third, related statement, this time from the Koran. A trivial first possibility can be ruled out immediately: The often (if rashly) claimed defeat of rationalism in the 1927 Copenhagen interpretation of quantum mechanics is not a door through which allegedly "irrational" statements like the three messages quoted above could seep in to get scientific acceptance. Esoterics is an equal enemy of both religion and science. A second, long-accepted functional interpretation proceeds along similar lines but is equally misleading: the allegation that the ethical education of the masses justifies "metaphysical myths" to be installed as lubricants for a better functioning of society.

The third—the only noble—alternative would be to check whether or not Locker's conjecture is sound. Three elements of modern science will be selected to test the claim: (1) The scientific status of the Now, (2) the biological phenomenon of mothering, and (3) the physics of Everett's many worlds.

THE GONG OF THE NOW

The Now has no place in physics so far—only simultaneity has. The Now is but a local parameter that can take on any value in principle. Roger Shepard, of Stanford University, maintains the Now is "private."[1] Unlike color, though, the Now is not an alien to the physical world. It is reserved a formal place in it and is, in this sense, not "subjective" despite the fact that there exists no machine to measure it. On the other hand, the Now shares with color and scent the property of "assignment."

Subjectivity, color, bodily identity and health are "assigned" and so is the whole universe with its laws and initial conditions. Still, if one takes the universe and all that is contained in it for granted, the meaning of the notion of assignment is not yet exhausted. "This given moment" (it is now ten minutes past three P.M. on December 16, 2004) is an assigned given. The Now is a typical assignment condition in the scientific sense. A change in the Now is worth the difference between life and death, between happiness and bereavement. It is about as incisive as if we could switch identities between different bodies. McTaggart[2] distinguished between one's being in the A-series or in the B-series of the scientific description of temporal evolution. His theorem that a formal incompatibility exists between the two series (between the Now-centered A series

---

1. In a personal communication in 1994.
2. McTaggart-Ellis, "The unreality of time," 457–74.

with its constant shifts and the invariant series) is still neither widely known nor, it must be admitted, fully understood.[3] A time machine like Gödel's,[4] supposing it were possible, would in effect allow you to change assignment. A change in assignment is a change of the world. Our being continually pushed along involuntarily as if sitting on the transparent seat of a ski-lift (by now it is already 3:19 P.M.) is so incisive an interference, perpetrated on us, that the question of its origins and its acceptability-in-principle poses itself. Descartes was one of the first to scientifically address this "question of the chalice," as it may be called.[5] This was in the year his young daughter had died at the age of five.

The notion of assignment conditions is a scientific one. The reason they are not much talked about may have to do with the fact that one cannot do much about them. Nevertheless, their sheer reality clearly calls for an answer to the question of their acceptability-in-principle and meaning.

BONDING

There is a book by Pfannek on the "Unmothering of the Soul" in the age of abortion.[6] The story of Isaac, the almost-sacrificed child in the Bible, contains two summing-up phrases in the form of names of places that can still be visited in the area in question. One is called "The-Lord-Is-Watching" (presaging, if you so wish, Orwell's phrase "Big Brother Is Watching," but so in an entirely benevolent fashion). The other, mentioned only a few lines later, is—in correct translation of the chosen tense—called "The-Lord-Can-Constantly-Be-Seen" (Gen 22:14).

Unlike the former statement which can only be believed or not, the latter is falsifiable since it makes a claim about something accessible in the real world. Is it true that the assignment-giving instance (A.G.I.) of science can be watched in action in its mothering role?

Let me take you on a little round-about tour to show that there exists a case in point in science. Konrad Lorenz got his "Nobel Prize" for the discovery of "imprinting" in young goslings. Nature has designed them in such a way that they must follow—desperately, "with all their might"—the first sufficiently large, sound-emitting moving object en-

---

3. Franck, "Virtual time," 57–81.
4. Gödel, "Cosmological solutions," 447–50.
5. Descartes, *Meditations*.
6. Pfannek am Brunnen, *Die Entmutterung*.

countered after hatching (for it most likely is their mother whom they need for survival). By going through those movements of desperately trying to follow a certain object, they get "imprinted" on that object, that is, get attached to it emotionally for the rest of their lives. The extreme malleability (at first) and extreme rigidity (subsequently) are the aspects which gave the phenomenon its name. But the more interesting aspect is the intensity of the "bonding" that develops. Bonding is a characteristic of all brood-rearing animals and, in particular, mammals—humans not excepted.

René Spitz discovered "hospitalism"—the predictable deaths of young children placed—because of some bodily ailment—into an optimally run hospital that lacked only one thing: a non-changing person to attach to and rely on—a mothering figure.[7] In our discussing the phenomenon, Lorenz and I discussed the van Hooff effect six years before van Hooff[8] described it independently: the fact that smile and laughter look identical in but one primate species—you probably guess which species that may be. Smiling, however, is the sign and signal of bonding.

Van Hooff used the words "smile" and "laughter" in the title of his paper[9] while in the text, the technical terms "silent bared-teeth display" and "wide-open mouth display" were defined and used exclusively—as befits a comparative-anatomical and physiological scientific context. We here return to plain language and, if you so wish, a journalistic style at the risk of van Hooff himself disproving of this procedure (he is a very conscientious scientist and, as he told me in 1973, abhors any functional conclusions to be drawn from his findings).

A misjudgment on the part of the young is the foreseeable consequence when not only the displayed happiness of the offspring is a rewarding signal for the feeding adult (as is the rule with highly social mammals like wolves), but also the displayed happiness of the adult is a rewarding signal for the offspring.[10] For a mere laughter (jolly happiness) has paradoxically the same rewarding effect on the child as a sign of genuine affection would have. Do the resulting erroneously triggered bonding bouts deserve the label "misjudgment" under all circumstances? This is the question that poses itself in the present context.

---

7. Spitz, "Hospitalism," 113–17.
8. van Hooff, *Phylogeny of laughter,* 209–41.
9. Ibid.
10. Rössler, *On the Animal-Man Problem,* 529–32.

As every parent knows, the human offspring starts to feed the human caretaker at a certain young age. This phenomenon is not much emphasized in the literature because it is so embarrassing from a biological point of view. Imagine a species in which the young start feeding the adults! Biologists are appalled at such a dysfunctionality judged from the point of view of the accepted theory of selection. They are used to speaking of a "lethal factor" for the long-term survival of the species in similar cases—like the preference of ultralong antlers by the choosy females of a certain species of deer. The only biological "excuse" in the present case could be that the behavior in question to some extent anticipates the end point of cosmic evolution in Teilhard's scientifically accepted asymptotic "point Omega," but this is a topic that cannot be addressed here.[11]

Is what happens in the playroom indeed a misunderstanding on the level of the individuals? As it turns out, the misunderstanding is no misunderstanding at all when it occurs symmetrically.

The phenomenon is easier to understand if we switch from humans to wolves. They too express both happiness and bonding by the same social signal—tail wagging, as every child knows from his pet. But in the canids, no similar misunderstanding occurs as occurs in humans. This is because the wolf lacks a certain capability of the brain that can be called "mirror competence." Not only do all humanoid apes, including gibbons, possess this trait, but so too do dolphins and their relatives and some further species like some corvids (and perhaps elephants). The trait refers to the ability to mentally rotate your environment precisely enough to discover the presence of an "exchange symmetry." This mathematical notion was first introduced by quantum physicist Wolfgang Pauli in the context of the behavior of indistinguishable particles like electrons. In our context, it applies to the fact that "giving" and "taking" are the same act if you exchange the identities of the two partners.

Since the happiness of either one ignites a "smiling feedback" if they are strongly bonding, it is no wonder that they soon discover—both Mom and toddler—that the good taste of an apple (say) makes both incredibly happy, no matter which one of them tastes and eats it. In other words, the suspicion of benevolence hovering over there on the other side, and the invention of benevolence on this side, inevitably arise in the toddler in his transactions with his bonding partner.

One could say that this is a misunderstanding much like the quoted claim of Jesus which ignited Locker's phantasy. In the present case, it

11. Cf. Rössler, *Nonlinear dynamics*, 47–67.

certainly is. For it indeed never was the happiness of the other side, only the degree of bonding-charm displayed by it, which triggered the toddler's own happiness in the interaction. But the impression of an identity under an exchange of positions between the sequentially occurring giving and taking motions on the two sides, enables the prediction of the presence of a "feeling soul" over there, both when it takes and when it subsequently gives.

As a consequence, the whole evolutionarily optimized machinery of autistic self-gratification (as it is needed for evolution to go on in the Darwinian paradigm) breaks down—in the greatest transformation of functioning conceivable and currently known. A no longer egocentric, but benevolent and benevolence-appreciating, "person" is projected to exist on the other side and, by this very attribution, hatched—on this side, the side of the toddler. This interactional process is the only example of a creation out of nothing known in science so far.

The mathematics of this process is too hard to be clearly deciphered by the brain of an adult person, Konrad Lorenz always excused himself. But of course, it is not too hard to invent and pursue to the end for a toddler.

One way to continue from here would be to go over to the theory of indistinguishable particles in both quantum and classical mechanics.[12] The other is to go into the theory of benevolence or—as Plato called it—the "Sun of the Good."

THE SUN OF THE GOOD

Modern, post-Bronze Age, society is constantly bewildered by the wave of sainthood that emerges with every new generation of toddlers every three years. Still, the phenomenon is not a topic that the public is fully aware of.

This may have to do with the fact that this naive stage does not last long in its pure form. The former toddlers at age four start to become quite proficient at lying—about twenty times a day, much like an adult (according to a recent estimate that made it into the newspapers of the world). Indeed, too much benevolence and trust is considered a sign of "dimwittedness" in society. At the same time, fathers admit to harboring a nostalgic feeling toward the age in question of their offspring. The same positive attitude is maintained by society toward people who by fate of

---

12. Rössler, *Jumping identities*, 307–19.

nature are forced to retain this early stage of full personhood, unbroken. Also, the guest-friend is traditionally in the same, both overprivileged and underprivileged, role.

The idea of benevolence is so foreign to modern post-medieval society that even science appears never to have touched on the origins of this phenomenon. (On *Google* not long ago, "benevolence theory" brought a mere dozen entries—and none of them in the context of science, all in the context of religion.) There is indeed no scientific theory of motherhood yet—of the motherhood of the soul that distinguishes the human species. Only some individuals like Gandhi and Mother Teresa got the label "-chi" attached to their name by the Indian public (which corresponds to the German suffix "-chen" for cute and little), in recognition of their archaic stupidity or not-so-stupidity, perhaps. These two persons, along with living scientist Luc Montagnier discoverer of the AIDS virus, enjoy an almost mythical status.

THE ASSIGNMENT OF WORLDS IN EVERETT'S THEORY

The most memorable episode in the recent biography of Everett, by Shikhovtsev,[13] is perhaps the sad story of his taking a hitchhiker along during a long cross-country car ride; the topic turned to religion and Everett later learned that owing to their conversation, the hitchhiker had subsequently committed suicide.[14] Everett's theory is very metaphysical, although in reality not frightening. Destructive misunderstandings must be prevented, however. I once had a student who invented the idea of an Everettian Russian roulette—to play Schrödinger's cat with a 1/6 instead of a 1/2 probability for some twenty rounds. He fortunately realized this was not the splendid idea it looked like at first sight. This is because even though an Everett world can never be exited, the hoped-for healthy last conscious moment before the bullet does its work (so that it could not have any destructive effect if that world survived) need not be the last one that is experienced—and hence the trick will not work. The combined scientific and religious experiment thus does not work.

The similarity to Loyola's self-experiments is striking but misleading, too. For example, when the founder of the first "society" (who called themselves friends of Jesus) wanted to visit the holy land, being a beggar-monk he could not offer any money to the captain of a departing

---

13. Shikhovtsev, *Hugh Everett III*.
14. Eugene Shikhovtsev, personal communication in 2002.

ship. Five times he tried. Why should I take you along?, he was asked each time. His answer: Because then you will arrive safely. Only the fifth captain agreed on the deal, and you can guess which ship was the one to arrive. Such experiments were called "exercises" at the time. However, this still is close to fundamentalism in all its dangerous forms.

Everett was not a fundamentalist, only a fountain of scientific ideas. He was Wheeler's (the inventor of the name black hole) brightest student. Whenever we observe a quantum decision that with the same probability could also *not* occur—like a Geiger-counter clicking or not clicking during the next ten seconds, when one knows that there are ten clicks per 100 seconds on the average—the world splits according to Everett into two worlds,[15] with no click occurring in the one and a click occurring in the other. You cannot predict in which world you awake.

Most scientists are reluctant to embrace Everett's theory in spite of its accepted mathematical correctness as a description of all known quantum phenomena, for they do not know that his many "worlds" are nothing but cuts (or interfaces) that run through the same single universe.[16] Thus his theory is actually nothing but an elaboration on the notion of the Now (to which we herewith return). While the common-sense Now is a point on a line, the new generalized Now discovered by Everett is a point on a plane. (Technically speaking, the Now thereby becomes "co-dimension-two" rather than co-dimension-one.) This idea that we ourselves, and all other things, change invisibly as time goes by, was already presaged by Heraclitus (so, for example, in his famous saying "Into the same rivers we step and step not, we are the same and we are not").[17]

The Now is not the only example of manifest assignment—if it is indeed true that Everett's theory can be experimentally distinguished from the Copenhagen theory (by a confirmation of the Bell correlations also when the two measuring stations are causally decoupled by their being placed into two mutually receding spaceships—one the earth, the other a satellite passing overhead).[18] The fact that no believing quantum physicist has any doubts about a positive outcome (at the expense of Copenhagen) may help explain why the Feingold-Shimony

15. Everett, *"Relative-state" formulation*, 454–62.
16. Rössler, *Relative-state theory*, 845–52.
17. Rössler, *Endophysics*.
18. Rössler, "Einstein completion," 367–73.

experiment—as it is called after its first cautious proponents—has still not been performed after almost fifteen years; a proposal to ESA is four years old (Anton Zeilinger, personal communication 2001).

The "Everett cut" is the second example of manifest assignment, it appears. While the assignment of a body, or a gender, or a race, or a species, or a century, is considered a "trivial assignment" since it is coarse-grained—and hence is believed to be "safe" in the sense that one needs not worry that it may suddenly be changed (although this evidently is naive)—the infinitely sharp micro-assignments of the "Now" and the "Cut" make the potential unfairness of it all palpable to a Western mind. It is like being talked to individually rather than as a group or an epoch (behind which one could hide). In this sense, Heraclitus' saying acquires a new ring: "The [One who controls everything including] War is the father of all: the ones he turns into gods and the others into mortals, the ones into free men, the others into slaves." Could it be that the underlying razor is micro-sharp?

Everett reluctantly worked for the military because he never found an academic assignment. Possibly, his superiors tolerated his working on a world-change machine, although no records of this can be found in the open literature. The latter would not—like the invention of the wheel or the bomb—change something within the world, but rather change the world as a whole. One could call it the "world bomb," although it would be intended by its designer to be only a "bombe glacée" (giant ice cream serving). However, in spite of the grandiose purpose, the machine would look rather ridiculous to a casual onlooker, because it would have the form of a helmet into which low-intensity microwaves of a tunable infrared frequency would be injected.[19] Everybody sees at a glance that such a device is ridiculous because it could not change the world in any other way than a drug would. Unexpectedly, this is not correct. If it were possible to manipulate the micro-interface (in between the consciousness-carrying particles in the brain and the rest of the universe), the whole interface, being the world to that person, would change objectively. So for the assigné; but so also for everyone else in his or her world.

The counterargument by the onlooker would be that even if this changes *your* whole world and all your partners within it, it will not affect mine if I decide not to buy a similar helmet. Again, a touch of

---

19. Rössler and Weibel, *Is Physics an Observer-Private Phenomenon Like Consciousness?* 443–53.

dimwittedness makes itself felt if someone should be naive enough to don the helmet. Is the universe paradise or is it not?

DISCUSSION

Three at first sight mutually unrelated strands were woven together: Now-assignment, benevolence-attribution, and world-assignment. The Sun of the Good has, mistakenly or not, been re-invented by every walking human person at a young age, if the interpretation of toddlerhood, given above, is not too naive an implication of chaos theory. Was the adoption of the Sun of the Good a misunderstanding? Or is the modern understanding of the Ten Commandments (as ten threats) perhaps mistaken? In an age before the broad invasion of the lie, the same Ten Commandments would perhaps have been just ten friendly reminders of how smoothly things work out if paradise is kept clean of cruelty. In some religious rituals, the Now of a certain moment in time (like the beginning of Sabbath) also still plays a role. It would act as a reminder that the whole world is a conscious, benevolent assignment of a private and intimate character.

This strange thinking was, in modern science, re-invented by eighteenth century Jesuit priest-cum-mathematician Roger Joseph Boscovich.[20] He called it the "principle of the difference" (if it is allowed to condense what he wrote into a single catchword; an English translation from the Latin is to be found in reference [14]). In 1905, famous physicist Lord Kelvin still could say of himself "I'm a Boscovichian pure and simple." And nineteenth century physicist James Clerk Maxwell invented, in Boscovich's footsteps, his well-known "demon" that is outside the universe and hence can see and do things that are impossible for any inhabitant (like violate the second law of thermodynamics). "Endophysics," the science of the universe as it looks from the inside, is a demonological discipline. It enlarges the traditional scientific program by giving one the freedom to put into focus and manipulate the interface. Nevertheless, assignment will always remain stronger than any freedom to manipulate the interface from within.

The art community has already responded with the creation of a full-fledged "endoaestetics" which began with the Gödel-inspired

---

20. Boscovich, *On space and time*.

installations of Peter Weibel of the 1970's.[21] Or is it the archaic (Shamanistic) triad of art, religion, and science that is trying to come back on a higher level?

How could people like Jesus and Buddha (and Buber and Levinas and Gandhi and Mother Teresa) be as social as they were? They returned to the mother, the Sun of the Good. The preposterousness of kindness and of expecting kindness is not naive: It can be turned into an industry. (This fact would, in the mentality of our age, be the ultimate proof that the world is paradise.) Lampsacus, hometown of mankind on the Internet,[22] is "dimwitted" enough a proposal along these lines. Even the fourth pillar of the ancient triad, paradisiacal survival, would cease to be a mirage if Markus' idea is put into action.

SUMMARY

Is the world a benevolent assignment? Most everybody would say that this is not a question that science can even try to answer. Therefore it was left to a young system-theoretician-cum-theologian to pose this question. Some preliminary remarks concerning the frightening nature of the Now, the disbelieving happy invention of benevolence by a toddler, and the disbelieving frightened discovery of the many quantum worlds by Everett, were offered and placed into this new context. The science of endophysics and the morals of Lampsacus (hometown) appear to be compatible with the strange new proposal.[23]

---

21. Gianetti, *Ästhetik des Digitalen*.

22. Rössler, *Das Flammenschwert*.

23. Acknowledgments: Alfred Locker is responsible: he gently forced me to follow up on my promise to submit a longer paper. I thank George Lasker, Ayten Aydin, Ken Hiwaki, Jerzy Wojciechowski, Hugh Gash, Bill Graham, Greg Andonian, Wim Smit, Jaap Hiddinga, Don Rudin, Winfried Rudloff, Eberhard von Kitzing, Gasser Auda, Wiktor Adamkiewitz, Michele Malatesta, Wendell Wallach, Martha Bartter, John Hiller, Jerry Chandler, Robert Taormina, Peter Mbaeyi, Martin Buber, Gerhard Heieck, Anton Traum, Guido Göhler, Iradj Rahimi-Laridjani, Dietrich Hoffmann, Masaya Yamaguti, Ichiro Tsuda, George Kampis, Marcus Fix, Peter Weibel, Detlev Linke, Niels Birbaumer, Siegfried Zielinski, Jürgen Parisi, Bryce DeWitt, Mohamed ElNaschie, David Finkelstein, Rüdiger Tschacher, Artur Schmidt, Adolf Muschg, Michael Langer, Ken Ford, Daniel Dubois, Oswald Bayer, Gregor Nickel, Matthias Wächter, Hans Primas, Michael Michelitsch, Stanislaw Dziwisz and you, my favorite reader, for stimulating support. For J.O.R.

4

Personal Knowledge and the Inner Sciences[1]

MARTIN ZWICK

ABSTRACT

This paper conceptualizes spiritual disciplines as sciences. It uses this conceptualization to probe into the similarities and differences between modern science and religious tradition, and into the cultural significance and possible future impact of the "new religions." The paper draws upon the ideas of Michael Polanyi as a possible bridge between science and religion, and proposes that these ideas are relevant not only to the major Western religions, but to Eastern and non-mainstream Western religions as well. Imagining science as a spiritual path, or gnosis, would challenge an exclusivist understanding of scientific knowledge, and suggest the relevance of such knowledge to wisdom. Interpreting the spiritual disciplines as inner sciences might help strengthen and purify religious practice, and lead also to a critique of science and a new conception of its possibilities. These unconventional perspectives provide a novel basis for a dialogue between science and religion. However, since there are many differences between "inner" and "outer" research, the metaphor of spiritual disciplines as sciences is limited; if taken too literally, it will obscure more than it illuminates.

---

1. A talk given at the conference on "*Other Realities: New Religions and Revitalization Movements*," University of Nebraska, Lincoln, Nebraska, March 27–30, 1985.

INTRODUCTION

One significant but not widely appreciated impact of the "new religions" has been to reopen the question of the relation of religion to science. I speak of new religions in the sense defined by Needleman in his book by that title,[2] that is, I am referring primarily to Eastern teachings which have gained adherents and cultural influence in the West over the past two decades. To some degree, certain of these religious systems can be viewed as encompassing "sciences" with well-articulated theories and powerful technologies, and it is this particular perspective on these religious movements which I would like here to explore. It may well be that the most substantial possibility of a creative dialogue between religion and science lies not in the encounter of the mainstream Western religions with science—which, of course, has a long history—but in the contemporary meeting of Western science with Eastern religion, in this "grand titration," to use Needham's expressive chemical metaphor, of West and East which is occurring in America today.

Many of these "new religions" appearing on the America scene are, of course, nontraditional and new in the West but completely traditional and old in the East; for example, Zen Buddhism, Tibetan Buddhism, and various schools of Yoga. Still, in their modern Western incarnations, these traditions are being altered, and so the appellation of "new religions" is, in many instances, appropriate. Some of these religions are Western, but not mainstream, such as Sufism, or trace their origins to the Western esoteric tradition, such as Kabbala or Steiner's Anthroposophy. Some have Eastern and Western origins, and are essentially syncretic creations, such as the Arica school founded by Ichazo or the teachings of Rajneesh.

What would I like to explore in this paper is a conceptualization of spiritual disciplines (primarily but not exclusively Eastern) as sciences, and to use this conceptualization to probe into the similarities and differences between modern science and religious tradition, and into the cultural significance and possible future impact of the new religions.

Before pursuing these notions further, it should be clear that this idea only characterizes one aspect of these spiritual traditions, and not actually their most essential aspect. Moreover, what is being developed here is "ideal type" analysis; that is, an abstract construction built around

---

2. Needleman, *New Religions*.

a composite of features from different traditions. It is beyond the scope of the present exposition to establish the degree to which any particular religious system may be usefully analyzed with such an idealization. (But just to illustrate the point: this idealization is particularly appropriate for Arica and Transcendental Meditation). The purpose of attempting such an abstract construction is to express in coherent form those features of Eastern (and some Western) religious traditions which seem to have a "scientific" character. The analysis might be regarded also as normative for efforts seeking an underlying unity between scientific and religious knowing.

In discussing these ideas, especially in the latter half of this paper, I will draw upon the ideas of the philosopher, Michael Polanyi. Polanyi's thought has been extensively utilized and cited as a bridge between science and religion, but discussion of his ideas so far has focused mainly on the implications of his post-critical philosophy for *Christian* theology.[3] I would propose that his ideas are relevant to Eastern and nonmainstream Western religions as well, although Polanyi did not himself address these traditions to any significant extent.

Polanyi's distinction between "verification" and "validation" is useful to introduce the notion of the "inner sciences" by distinguishing between the relationship between modern science and the mainstream Western traditions as compared to that between modern science and Eastern (and some Western) traditions. Polanyi regarded religion as offering, as does science, a systematic mode of ordering our experience, and considered that a process of a "validation" in religious experience was analogous to the process of "verification" in science.

> The acceptance of different kinds of articulate systems as mental dwelling places is arrived at by a process of gradual appreciation, and all these acceptances depend to some extent on the content of relevant experiences; but the bearing of natural science on facts is much more specific than that of mathematics, religion, or the various arts. It is justifiable, therefore, to speak of the verification of science by experience in the sense which would not apply to other articulate systems. The process by which other systems

---

3. See, for example, the papers thematically devoted to "science and religion in the thought of Michael Polanyi," in *Zygon* 17 (1982); Also "Science and Religion in the Thought of Michael Polanyi"; American Academy of Religion Conference, Chicago, 1975: "Towards a Post-Critical Theology: The Influence of Polanyi"; Richard Gelwick, "Polanyi-Tillich Dialogue of 1963," 11–19.

than science are tested and finally accepted may be called, by contrast, a process of validation."[4]

This distinction, between verification and validation seems useful for distinguishing between modern science and Western religion, but for the Eastern traditions, verification seems an appropriate concept, as we shall see, since these traditions can be viewed as sciences in which knowledge is tested in individual experience.

THE "INNER SCIENCES"

The starting point for this analysis is the observation that some Eastern and Western spiritual disciplines, such as Zen, Yoga, Sufism, Tibetan, Buddhism, etc., can be regarded as constituting "inner sciences," disciplines in which the experimenter, experimental materials, and apparatus, are all simply the individual him- or herself, and whose aim, at least in the initial stages, is the gaining of self-knowledge.

To be more specific: spiritual disciplines generally include inner experiments and exercises of various kinds. These include, for example, the vocal or silent utterance (and repetition) of sounds (mantras), internal visualizations or meditation upon external visual symbols (mandalas), gestures, postures, and/or movements (mudras, yogic positions, devotional or meditative dance forms), efforts to focus concentration or expand awareness or monitor thoughts, sensations, etc. Exercises may involve ordinary physical activity, such as manual labor, crafts, etc., and may extend to social behavior, i.e., to interactions which family, friends, coworkers, etc., and to relations with nature.

In so far as the researcher and the subject of research is the same, and thus knowledge is personal both in what it concerns and by whom it is used, these inner sciences differ from the science of psychology, in which the roles of researcher and subject are usually distinct and whose goal, like that of all the sciences, is the production of a publicly available store of general knowledge and technique, which benefits individuals only indirectly, after a gradual process of societal diffusion, assimilation, and application.

The effective performance of these exercises, like laboratory experiments, requires commitment, skill, and understanding. Conditions for performing these requirements and results obtained may be, as Charles

---

4. Polanyi, *Personal Knowledge*, 202.

Tart has noted, "state specific," i.e., they depend on the state of the experimenter's body, feelings, mind, and consciousness, and his or her degree of moral and spiritual attainment.

Traditional yogic practice, for example, encompassed widespread investigation of internal sensations and the possibility of their manipulation. This entire field of exploration has stimulated investigation by the tools of modern science—the whole field of biofeedback research. While once the possibility of conscious control of autonomic processes was dismissed by science, the yogis long ago knew that this was possible.

More subtle and more difficult is experimentation which focuses on the relation of the individual to his/her external world, to personal and social interactions. Here research typically yields both new findings about oneself and the world, and the discovery of the incorrectness of previously held views. It has the taste of scientific activity most vividly in the shock of unanticipated, indeed, often unwelcome results, in the need to repeat experiments, and in the difficulty of dispelling, or even detecting, preconceived biases. The goal of these experiments is, as has been suggested, the acquisition of "personal knowledge," in the sense of Polanyi, and the sense of Socrates.

THEORY

The doctrines of these traditions often embody sophisticated theories for which the experimental results can provide confirmation. When these theories have some discernable structure, they often look like "systems theories," that is theories which emphasize the similarity of relationships between different phenomena, even between phenomena at different scales. Alchemy "(which has received little attention in the new religious movements, although the Jungians seem to have taken up its study) illustrates this clearly: there was presumed to be an isomorphism between external chemical and internal psychic phenomena. The Hermetic credo, "As above, so below," also illustrates this mode of thought. Chinese religious philosophy emphasizes "correlative tabulations," as Needham[5] has noted; these reflect a "philosophy of organism," which in contemporary terminology would be called systems theoretic. Many forms of number mysticism also illustrate this approach.

---

5. Needham, *Science and Civilization in China*.

In the inner sciences, theory is often veiled, or the exercises indirectly related to the doctrine, to counter tendencies towards suggestibility and thus promote genuine empirical study. Sometimes exercises may even be given which are actually impossible to accomplish; this promotes the integrity of research and deepens investigation. The objective is the experimental tasting of reality and not its theoretical formulation, and esotericism is one means used to promote this objective. The theory is scaffolding, not the actual building, which is a personal construction and the result of individual work. As an inner scientist, the mystic is an empiricist, and belief is only hypothesis yet untested. In this de-emphasis on, and in some instances, scorn for, intellectualizing and theory-building, the inner sciences differ sharply from conventional science, as well as from Western intellectual traditions which might seem to resemble Eastern thought, such as phenomenology, or "pre-scientific" introspectionist psychology.

SOCIOLOGY

In some traditions, the results of experimentation are discussed with others in, as it were, "scientific" meetings, at which individuals report their efforts and findings. Techniques are shared, and experiences confirmed by others, while the exchange energizes the participants, and suggests new ideas for study. Often the group undertakes a common line of investigation.

The relation of disciple to teacher resembles, in a way, the apprenticeship of a graduate student to a faculty researcher. The student independently pursues his or her own "project" with periodic counseling from the advisor, or often a senior assistant, who is engaged in similar study at a more advanced level. This is obviously only partial view of the teacher-student relationship within this kind of religious practice, but it does highlight some features actually present in such relationships, which are not normally associated with the more familiar image of guru and chela, or the Western model of therapist and patient.

In this research mode of spiritual guidance, little use is made of transference; indeed it is assiduously undermined. The identification of the student with the person of the teacher or the founder of the religious movement interferes in the student's capacity for independent and unbiased investigation. Even the ideas of the teacher finally become a stumbling block to further progress. The aphorism, *"do not follow in the footsteps of the ancients; seek what they sought"* expresses this understanding. Nonetheless, the student is ill advised to undertake independent experimentation before some significant measure of competence and judgment has been achieved. An initial period of apprenticeship is virtually always necessary.

And it is not uncommon for young monks to round out their training by going to other labs.

TECHNOLOGY

Implicit in the notion of an inner science is that of a corresponding technology, which supports the scientific endeavor, i.e., provides the tools of experimentation, and also develops as a result of it. (Arica, for example, once explicitly advertised itself as offering a "spiritual technology." Transcendental Meditation has something of a similar character.)

But this technology is fundamentally a "tacit" one (Polanyi), and here is a source of one of many differences between the inner and outer or conventional sciences. While it may appear that many spiritual "techniques" are quite specifiable, such as those involving sounds or postures, the efficacy of these methods depend upon an irreducible tacit or unspecifiable component, and as the researcher advances in his or her practice, the domain of the tacit grows. Yet this does not preclude communication between persons having common experiences; it merely requires that communication be subtle and skillful and that a certain intimate relationship exist between participating individuals.

In the inner sciences, too, technology can subvert. Subverting in this domain means the utilization of spiritual techniques for purposes foreign to the spiritual undertaking. For example, there exist techniques and exercises for the development of will; at the service of ego, these not only become obstacles for further development, but can actually have harmful effects on oneself or others. Gide's *Lafcadio's Adventures* can serve to illustrate this point clearly, though it is not at all about a spiritual journey. Actually, one might well regard this as a story of "black magic" (magic not serving the values of a spiritual undertaking). Huxley's *Grey Eminence* is a more direct example: Father Joseph, the "eminence grise" behind the power of Cardinal Richelieu, was—at least in Huxley's account—deeply involved in spiritual discipline, and this involvement gave him added strength and charisma which helped him launch the Thirty Years War, which cost very many lives. (France, Father Joseph was certain, was the instrument of God.)

In the Eastern traditions, the worship of technology for its own sake and for the personal benefits to be derived from it is warned against. Siddhis or (magical powers) are supposed to accrue to those who persist in their spiritual practice. Leaving aside completely the question of whether such powers actually exist, and are not metaphors for capacities that would not really challenge the Western scientific worldview, we can note that in the all the traditions, the seeker is warned against the search for, or the indulgence, or fascination in these siddhis.

We are familiar with the ways in which modern technology may be, and has actually been, abused. This occurs at a micro-level, at the level of the person (e.g., the habitual ingestion of powerful chemical agents), and at the macro-level at the level of society (e.g., the development of improved techniques for massive mutual annihilation in war). There are perhaps comparable abuses of inner technologies. The undisciplined and unsupervised (or wrongly supervised) seeker can destroy her or her psyche and personality. This is the negative aspect of some "cults" that so concern many of us. But there are also societal abuses and societal harm. Koestler's *The Lotus and the Robot*[6] might be regarded in part as an indictment of the harm brought to Indian culture by the excessive dominance of the philosophy and practice of yoga.

There is still a deep anomaly in the relationship between the inner sciences and the technologies associated with them. The very existence of such a distinction constitutes a contradiction and a barrier to the seeker. The learning of technique binds the learner to technique and to the striving for results from its application, a striving which is counterproductive. Krishnamurti, despite—or perhaps because of—his Indian heritage, disowns all techniques of meditation, such as mantras, as mind-stultifying rather than mind-liberating. One way of dealing with this contradiction is to embrace it and work with it; hence, perhaps, the ubiquitous use of paradox in many traditions, e.g., Zen. Still, the contradiction is unavoidable. The beginner needs technique and must be motivated at least in part by egoistic goals; indeed at all stages the seeker is so motivated. Ideally, there should be no separation between method and the spiritual undertaking which it serves, between inner science and inner technology, but one cannot start at this point. The Sufis prescribe this formula: "First quit the world. Then, quit quitting."

---

6. Koestler, *Lotus and the Robot*.

PERSONAL KNOWLEDGE

In ordinary knowledge, including that gained by scientific research, the "personal," Polanyi asserts, is other and more than the merely subjective. It is the embodiment of the universal, in so far as it entails a dedication to truth, independent of its advantage or disadvantage to its recipient. As Polanyi notes:

> [W]e may distinguish between the personal in us, which actively enters into our commitments, and our subjective states, in which we merely endure our feelings. The distinction establishes the conception of the personal, which is neither subjective nor objective. In so far as the personal submits to requirements acknowledged by itself as independent of itself, it is not subjective; but in so far as it is an action guided by individual passions, it is not objective either. It transcends the disjunction between subjective and objective.[7]

Polanyi here calls attention to the fact that the scientist submits internally to accepted "scientific standards for the appraisal and guidance of his efforts," yet is simultaneously guided by his or her own hopes, expectations, or curiosity. The subjective side affirms an idea or theory; the objective side denies, i.e., imposes constraints of acceptability. Or from another perspective, it is the reverse: the world of science gives the impetus, and defines the problem, and it is the subjective world of the scientist which accepts the challenge and provides the resources. The personal is the synthesis of the two, the ground on which they are joined. In this dialectic is located much of the deep spiritual value in science.

To the extent that the submission referred to earlier is not merely an introjection of the standards of the scientific community and internalization of the requirements of professional survival, to the extent that the acceptance of objective standards is assimilated into the being of the scientist as a respect for, even a love of, the truth, to this degree is science truly ennobled by its aspirations to objectivity. This submission is a real phenomenon. It is a personal accomplishment whose presence or absence is often discernable in the training of young scientists, who differ greatly in the degree to which enthusiasm is successfully blended with a critical faculty. One can detect in many scientists an attitude of genuine impartiality, very close to a spirit of non-attachment. However,

---

7. Polanyi, *Personal Knowledge*, 300.

this achievement is not common, and is usually partial, influencing only that part of the scientist's personality engaged in professional or intellectual matters.

It is an achievement which is respected within the community, but which is not actually necessary to the success of the scientific enterprise, whose objectivity and hence progressive nature is guaranteed primarily by automatic social mechanisms.

There is a similar relationship between the subjective, the objective, and the personal in the inner sciences, although the situation is more complicated. The subjective is in the seeking of the student and willingness to experiment, in the resources for self-study, and most especially in the subject matter itself. The objective to which the individual must submit, in Buddhist terms, for example, is the Buddha (which may represent the historical person of the original teacher or have a deeper metaphysical meaning), the Dharma (the teachings), and the Sangha (the community of seekers).

It is in the personal that these two worlds are joined; that the mysterious reconciliation of the unique and the universal is accomplished. The personal characteristics of spiritual teachers vary considerably, but the Dharma is one. There is a Hasidic story retold by Buber about a rabbi who was asked by his disciples why he did not follow the practices of this own former teacher. Puzzled, the rabbi asserted that he did indeed follow his teacher, but this did not satisfy his students, who protested that in this or that matter the rabbi departed from earlier practice. Finally, the rabbi settled the matter: "I follow the Master exactly. He did not imitate, and I do not imitate."[8] The universal must be manifested in the personal, via a synthesis which bears the idiosyncratic stamp of an individual who struggled and came to understand. Hence, the aphorism, "beware the guru who does not have his own doctrine." There is both glory and tragedy in this achievement: The contribution of the subjective is necessary, but much misunderstanding and error flows from it. The paradoxical interdependence of universal and personal knowledge is apparent not only in the figures of the great teachers. Even a beginner on the way has some experience of it.

A commitment to objectivity is the hallmark of the personal. In the inner sciences, the field of experimentation covers one's entire life and thus objectivity is more difficult, and the ability to be objective cannot be

---

8. Buber, *Tales of the Hasidim*, 157.

assumed. Indeed it is a goal of training. Objectivity is not only a value; it is actually a spiritual "power." It is not a faculty of thought, but a context in which thought can be received. It is a capacity which applies as well to emotion and sensation, and does not diminish, but rather intensifies them. Meditation might be regarded as training to be impartial, and the development of an internal "witness" is an important preliminary accomplishment in a number of traditions. The difficulty of being objective about anything is valuable knowledge which spiritual teachings try to make accessible to their followers.

It is perhaps ironic that in the 1970's countercultural critiques of science should attribute its dehumanizing and alienating effects in our culture to its aspirations towards "objective consciousness," when in fact, objectivity is a value held in common by both modern science and spiritual tradition, and is one basis for a meaningful dialogue between the two. Of course, what writers such as Roszak[9] are attacking is the devaluation of personal experience, which seems so salient a feature of the prevailing ideology of science; and here Polanyi stands with the critics in his rejection of positivism and in his insistence that our conception of science must reflect the fact that knowledge is intrinsically personal.

But in this attempt to fashion a human-centered image of science, Polanyi has prepared the ground for a new and deeper criticism of science, one which I do not assume he would have supported. If knowledge to be meaningful must be personal, we may ask if this condition is sufficient or merely necessary. How meaningful, actually, is this knowledge of the scientist? As the possession of the social collectivity, it is certainly "meaningful"; at least in the sense that it has affected every aspect of human life. But what is its significance to the scientist?

And how does it compare to the knowledge gained in the "inner sciences"? We might ponder the words of Don Juan to Castaneda, after the two have just met, and the latter is condescendingly granting the status of equal to the uneducated Indian with whom he is speaking. Don Juan says to the anthropologist that he is a pimp, because the knowledge he gathers is not for himself, but for others. The quarrel here is not with the pursuit of knowledge to better the human condition. The deep achievement claimed for science is rather that it brings us to a fuller understanding of the universe. Don Juan denies this.[10]

9. Roszak, *Where the Wasteland Ends*.
10. Casteneda, *Journey to Ixtlan*, 81.

Scientific knowledge, while having a personal component does not, by comparison with the knowledge of the yogi (or sorcerer), belong to the scientist. It is knowledge which is only slightly "embodied," and barely tasted, hence (or, one reason for) the insatiable appetite for new research fields, new discoveries, as if some new knowledge might finally satisfy. From the point of view of the meditational disciplines, the knowledge of science serves mainly ego, and can be assimilated only by thought, leaving us in fragmented relation with our feelings and sensations.

There are certainly exceptions. I remember being almost startled by the words of a scientist friend and teacher of mine who, in a discussion of "mysticism," said to me (I paraphrase): "I do not need it. When I fly on an airplane, I understand how it flies, and when I see a blade of grass, I understand also something of its function. So I feel in relation to the world, and not apart from it."

But for the majority of scientists, the personal element of scientific knowledge does not penetrate widely into other spheres of individual experience, or deeply into being, and so remains unrelated to the whole, and therefore alienated.

While personal knowledge is necessary for scientific creativity and is the embodiment of scientific understanding, it is not, as in the inner sciences, intrinsically valued as a source of personal development. That a scientific career might contribute to the unfolding of wisdom and the purification of character is an unfamiliar proposition. Certainly, the organization of scientific activity is not directed at such aims. But could it be? Could one imagine science as a sadhanna, or spiritual path? Can one conceive of science returning, as Roszak urges, to "gnosis" as its framework and purpose?

These are questions raised by a consideration of the personal element in science, and by an interpretation of the spiritual disciplines as inner sciences. From this juxtaposition, we can derive a critique of science and a new conception of its possibilities. We also find the basis for a dialogue between science and religion, of a kind different from the interaction to which we have become accustomed, and perhaps also an entrée into Eastern traditions that is congenial to temperaments molded by a science-dominated culture. Indeed, it may well be the scientific aspects of some of the "new religions" which has made them so attractive to westerners.

But on this last suggestion, and before concluding this essay, a cautionary note, perhaps even a disclaimer, must be repeated. The conception of spiritual disciplines as sciences is a limited perspective, which illuminates some aspects of these traditions, and might even contribute to a strengthening and purification of religious practice. But there are many more major differences between the "inner" and "outer" sciences than have here been noted. The metaphor is limited, and, if taken too literally, it will assuredly obscure more than it illuminates. It is a conception which should probably be discarded as soon as it is grasped.

5

Symbolic Structures as Systems

On the Near Isomorphism of Two Religious Symbols[1]

MARTIN ZWICK

ABSTRACT

Many symbolic structures used in religious and philosophical traditions are composed of "elements" and relations between elements. Similarities between such structures can be described using the systems theoretic idea of "isomorphism." This paper demonstrates the existence of a near isomorphism between two symbolic structures: the Diagram of the Supreme Pole of Song Neo-Confucianism and the Kabbalistic Tree of medieval Jewish mysticism. The similarities of these two symbols in form and meaning are remarkable in the light of the many differences that exist between Chinese and Judaic thought. Intercultural influence might account for these similarities, but there is no historical evidence for such influence. An alternative explanation would invoke the ubiquity of ideas about hierarchy, polarity, and macrocosm-microcosm parallelism, but this does not adequately account for the extent of similarity of the symbols. The question of how to explain their resemblance remains unresolved.

---

1. A shorter version of this paper has appeared as "The Diagram of the Supreme Pole and the Kabbalistic Tree: On the Similarity of Two Symbolic Structures," *Religion East & West, the Journal of the Institute for World Religions*, 9 (2009) 67–87.

INTRODUCTION

A "SYSTEM" IS A set of elements and relations between elements. Two systems are isomorphic if the elements of one can be mapped onto the elements of the other with the same relations holding between corresponding elements. Symbolic structures are systems, and this paper notes a near isomorphism between the structures of two religious-philosophical symbols: the Diagram of the Supreme Pole[2] (*Taiji tu*) of the Chinese Song Neo-Confucian School (11th and 12th century) and the Kabbalistic Tree of the medieval Jewish mystical tradition (Figure 5.1). The elements and the relations between elements in the Diagram of the Supreme Pole (referred to henceforth as "the Diagram") can be mapped onto the elements and the relations between elements in the Kabbalistic Tree (referred to henceforth as "the Tree"), and when this is done many of the corresponding relations are similar. While corresponding elements differ in meaning due to differences between Chinese and Jewish thought, their roles within their respective structures often resemble one another.

The idea of isomorphism is relevant not only for comparing different symbolic structures but for describing the use of such symbols. Chinese thought correlated many phenomena with the Two Forces (Yin and Yang) or with the Five Agents (Earth, Wood, Metal, Fire, and Water) and similar tabulations were ubiquitous in European pre-scientific writings, including those of the Kabbalah. "Correlative tabulations" are implicit—and inexact—isomorphisms. Needham called such tabulations "proto-scientific,"[3] and one might more specifically regard them as an early form of systems thinking. Modern systems theory revives this analogical mode of thought but formalizes it. Instead of tabulations justified by intuition, relations are defined mathematically. If the same relations hold between corresponding elements of two systems, the systems are mathematically isomorphic.

---

2. The major alternative translation is "Supreme Ultimate." Needham's (*Science and Civilization in China*) translation of the word as "Pole" is used in this paper, despite the fact that "Ultimate" is more common. See also footnote 31.

3. Needham, *Science and Civilization in China*, 97.

FIGURE 5.1. Diagram of the Supreme Pole (left) and the Kabbalistic Tree (right).[4]

The numerals, I—V, label Diagram components (substructures), not individual elements, e.g., II includes the Two Forces (Yang and Yin); III includes the Five Agents (Fire, Water, Earth, Wood, Metal). The structures correspond if either one is left-right reversed.

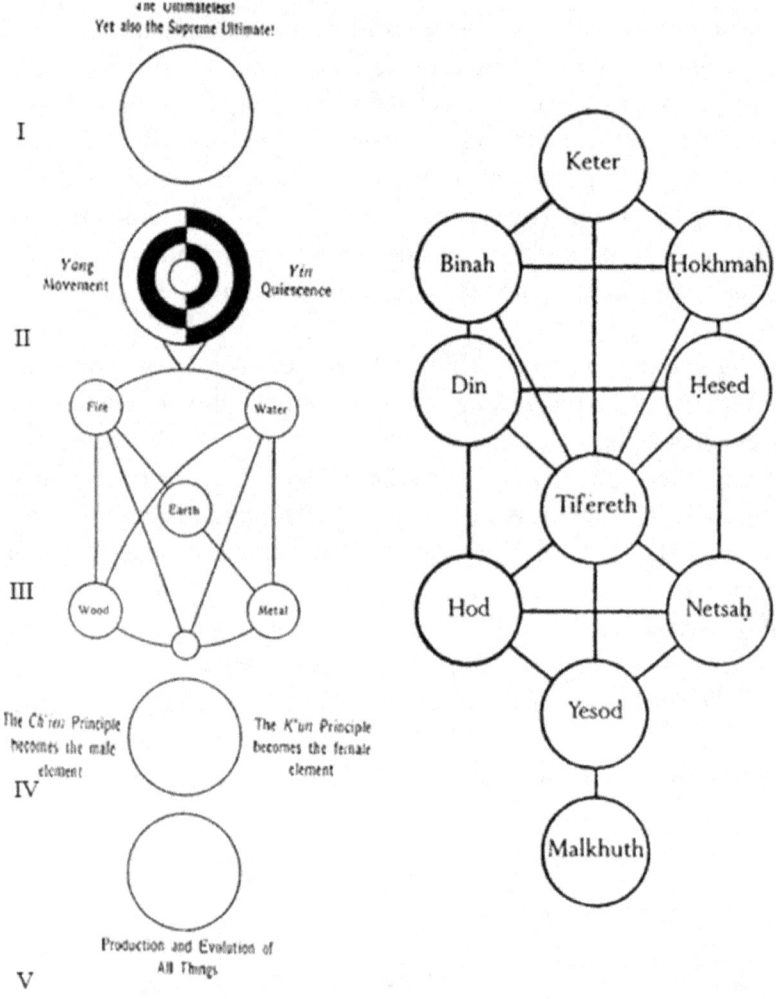

4. The Kabbalist Tree is from Scholem, *On the Mystical Shape of the Godhead*, 44; the Diagram of the Supreme Pole is from Fung, *History of Chinese Philosophy*, 436.

A classic illustration of isomorphism is the analogy that exists between electrical and mechanical systems, in which variables and parameters of one system type map onto those of the other type, and these elements are related in both via a 2nd order differential equation.[5] Mathematics not only makes the analogy exact; it also defines the limits of its scope. (The electrical system and the mechanical system differ in aspects not included in the isomorphism; for example, only the former can give electric shocks; only the latter manifests visible motion.)

This electrical-mechanical isomorphism is quantitative, but an isomorphism can instead be qualitative. For example, two systems might have the same graph-theoretic structure. The isomorphism would then consist in the existence of links (relations) between corresponding elements in the two systems, where the nature of these relations need not be specified. For example, if system1 has elements A, B, and C and links AB and BC,[6] and system2 has elements D, E, and F, and links DF and FE, then by mapping A onto D, B onto F, and C onto E, the relations are preserved, i.e., AB maps onto DF and BC maps onto FE, and the two systems are isomorphic.

---

5. The electrical system contains a resistance, capacitance, inductance, and applied voltage; the mechanical system is a disk that rotates in a dissipative medium and is connected to a spring that also resists the rotation. The correspondences are:

	(i)	(ii)	(iii)	(iv)	(v)	(vi)
Electrical	charge	Current	voltage	inductance	resistance	capacitance
Mechanical	rotational angle	rotational velocity	torque	moment of inertia	rotational resistance	rotational spring constant

Both systems obey a differential equation of the form, $a\, d^2x/dt^2 + b\, dx/dt + c\, x = e$, where x = (i), dx/dt = (ii), and e = (iii), and a, b, and c depend on (iv), (v), and (vi), respectively.

6. In simple graphs, a link connects only two elements, but links need not be dyadic. For example, in the graph-theoretic structures used in Reconstructability Analysis (see Zwick, "Overview of Reconstructability Analysis," 877–905), triadic, tetradic, etc. links (relations) are also possible between elements. (Graphs that have such relations are "hyper-graphs.") In principle, symbolic structures could exhibit such higher ordinality relations between their elements, but usually only pair-wise relations are considered. One analysis of symbolic structures that begins to explore higher ordinality relations is J. G. Bennett's "systematics" (not to be confused with the word's meaning in biological taxonomy) (Bennett, *The Dramatic Universe*). The syntactic (but not semantic) aspects of Bennett's framework of number symbolism has close affinity to graph-theoretic analysis of structure, and systematics can be thought of as the Reconstructability Analysis of ideas, as opposed to quantitative data.

The similarity of the Diagram and the Tree is graph-theoretic; there are no quantities that might be related by some differential equation. But the symbols are plainly not completely isomorphic. For example, the Diagram is "partially decomposable"[7] into separate components I to V, while the Tree is a single connected graph. What is especially similar in these symbols is the relative *spatial* arrangement of the elements, i.e., their vertical and horizontal locations, more than their specific connectivities. The Diagram and Tree both make use of a "dimensional domain"[8] in which elements are organized vertically by the principle of hierarchy and horizontally by the principle of polarity. The symbols are thus more than graph-theoretic structures: relations between elements are defined not only by connectivity but also by spatial location.

CHRONOLOGY, OVERVIEW, AND SOURCES

Since the most plausible null hypothesis about a cosmological symbol from Neo-Confucianism and a theosophical symbol from Kabbalah is difference, similarities are noteworthy, but differences are no less important, and one prominent difference between these two symbols is their status within their individual traditions. The Diagram had Daoist precursors[9] and its importance to Neo-Confucianism was evident at the inception of this movement. By contrast, the origins of the Tree are shrouded in mystery. As a canonical structure it appears late in the Kabbalist tradition, more as a visual mnemonic than as a symbolic centerpiece.

The symbols differ greatly in the precision with which their first appearances can be dated and the degree to which a few seminal writings gave them definitive interpretations. Two principal commentaries on the Diagram were written: one in 1060 by Zhou Dunyi, who recast an earlier Daoist symbol into Neo-Confucian form, and the other in 1175 by Zhu Xi, a later—and the most prominent—philosopher of the Song Neo-Confucian school. The emergence of this school is described by Fung as follows: "By the beginning of the Song Dynasty, i.e., around the year 1000, the major existing schools of thought (Confucianism, Taoism, and Buddhism) had all reached roughly comparable stages of develop-

---

7. Simon, *Sciences of the Artificial*.
8. Angyal, "Structure of Wholes," 25–37.
9. This is the dominant view and is assumed in this paper, but Robin R. Wang ("Zhou Dunyi's Diagram," 307–23) mentions an argument that the diagram was original to Zhou Dunyi and was plagiarized by a Daoist in the Song Dynasty.

ment in the course of which a considerable intermingling of ideas had occurred. All that was lacking was the series of great men who were presently to appear, and were to organize and unify all that had gone before into one great system."

Zhou Dunyi and Zhu Xi, among others, accomplished this unification. Driven by the desire for a coherent cosmology and by the syncretic motive of linking Confucianism to the other Chinese traditions, the Song scholars produced a Neo-Confucian metaphysics influenced by Daoism and Buddhism.[10] The Diagram of the Zhou Dunyi was the symbolic centerpiece of the Song Neo-Confucian synthesis.

By contrast, the Tree appears late and its origin is obscure. There is no definitive treatment of the symbol that is analogous to the two commentaries on the Diagram. The *Zohar* (ca. 1286, Moshe de Lèon, Guadalajara, Spain)was the central book of the Kabbalah, but Kabbalist doctrine had roots in many earlier works, including the *Sefer Yetsirah*, third to sixth centuries, and *Sefer Bahir*, 1150–1200, Provence, France.[11] The Tree did not appear in these books, emerging as a canonical structure only in the fourteenth century.[12] It was not a central symbol for the Kabbalists. The prominence it later gained is partially due to its importance in occult and Christian Kabbalah. It was the doctrine of the *Sefirot* (plural of *Sefirah*, literally "enumeration")—the ten elements of the Tree—that was central to the medieval Jewish mystical tradition. The *Sefirot* were religious concepts long before they were integrated and visually represented in the Tree. Similarly, the Chinese doctrines of the Two Forces and Five Agents predated their use in the Diagram.

The subjects of these symbols, although not the same, play similar roles in their respective cultural contexts: for the neo-Confucians, the fundamental metaphysical principle, the Supreme Pole, with its Forces, Agents, and other manifestations; for the Kabbalists, God, with the *Sefirot* representing divine attributes or instruments.[13] To the Western mind,

---

10. This is a paraphrase of the account of Henderson, chapter 4: Correlative Cosmology in the Neo-Confucian Tradition, *The Development and Decline*, 125.

11. The dates of these works are uncertain and in dispute. Dates given here are from Scholem in *Kabbalah*, 57 (for the *Zohar*), 27 (for *Sefer Yetsirah*), and 42 (for *Sefer Bahir*).

12. Scholem, *Kabbalah*, 106.

13. Idel (*Kabbalah: New Perspectives*, 137) distinguishes between this common view, (1) the *Sefirot* as the components of the "divine essence," and its variations, (2) the *Sefirot* as "nondivine in essence" but as "instruments" or "vessels for the divine influx," and

the Diagram is philosophical ("cosmological") while the Tree is religious ("theosophical"). One could say that the Diagram is also religious, just not in the Western sense of implying a personal, law-giving, creator God. Conversely, given that for the Kabbalists, the structure of God was mirrored in His creation, Kabbalah also offers a cosmology.[14] This emphasizes its Neo-Platonic aspects, but in Kabbalah, mythological and biblical aspects predominate, and these have no Chinese parallel. Nothing in the Diagram corresponds to applications of the *Sefirotic* doctrine to biblical persons, passages, and events, or the mystical aspects of the Hebrew language. The differences between Neo-Confucianism and Kabbalah, and between Chinese and Jewish thought are substantial. Given these differences, the similarities of the symbols are striking.

These symbols were not only cosmological or theosophical. Both Neo-Confucianism and Kabbalah asserted the parallelism of macrocosm and microcosm. For the Neo-Confucians, this is illustrated by Zhou Dunyi's use of cosmological ideas for moral discourse. His "It is man alone who receives the finest (substance)," is a dramatic application of cosmology to anthropology. The Confucian centrality of human action is reaffirmed, deepened by a new metaphysical foundation. A human focus also characterized the Daoist precursor of the Diagram, where it referred to the "subtle body" of man which was the instrument and object of meditation. Similarly, as Idel notes, Kabbalah was both theosophical and "ecstatic."[15] The *Sefirot* applied to the human body, psyche,[16] and be-

---

(3) the *Sefirot* as "divine emanations within created reality," i.e., as "the immanent element of divinity"

14. Scholem (*Kabbalah*, 96) explicitly rejects the view of Franck that the Kabbalah was pantheist, but it is not necessary to go this far to see a cosmology in Kabbalah.

15. The dichotomy of theosophical (theoretical) and ecstatic (experiential) Kabbalah corresponds to a predominant focus on macrocosm and microcosm, respectively, but there is a continuum from theosophy to prayer to meditation. Where to place the "mystical" along this continuum is not always clear. The psychological interpretation of the *Sefirot*—which merges with the meditational and mystical—is more identified with ecstatic Kabbalah (e.g., Abulafia); it was de-emphasized in Lurianic Kabbalah but was later extensively taken up in Hasidism (Idel, *Kabbalah: New Perspectives*, 148–50).

16. Idel (*Kabbalah: New Perspectives*, 152) remarks about the later Hasidic emphasis on the psychological interpretation of the *Sefirot*: "Thus, the entire zoharic and Lurianic superstructure is viewed, not only as comprised in man ... but, according to Rabbi David's testimony, only in man. According to the Hasidic sources I am familiar with, Kabbalah is preeminently a paradigm of the human psyche and man's activities rather than a theosophical system." The human-centeredness of traditional Judaism was reaffirmed in Hasidism, gaining vigor and subtlety from the powerful adventure of Kabbalah.

havior, and to meditative and mystical practice. In the doctrines of *Shi'ur Komah*, the measurement of the "bodily parts," as it were, of God, and *Adam Kadmon*, the primordial man or cosmic anthropos, the Kabbalists gave symbolic human physical form to God. The Diagram and Tree thus depict not only cosmos and God, respectively, but also human physical, moral, psychological, and spiritual structures. Both symbols were used to declare that by perfecting oneself, one harmonized the macrocosm.

The literatures relevant to these symbols are large and diverse. The Diagram was Confucian, but had Daoist origins, and showed Buddhist influence; the focus here is on the Confucian and Daoist sources. In addition to the original Jewish Kabbalah, there were Christian and occult offshoots, and Jewish Kabbalah gave much less emphasis to the Tree than these later derivatives. Even within Jewish Kabbalah there were various doctrines; this paper emphasizes early (pre-Lurianic) Kabbalah.

The scholarly literatures on Neo-Confucianism and Kabbalah also differ in the extent to which they are dominated by a single investigator. For Neo-Confucianism and the Diagram, this paper relies heavily on Needham and Fung, especially Needham, whose translations[17] of Zhou Dunyi's and Zhu Xi's commentaries are used in this paper. Unless otherwise noted, all references to these authors are to these translations, which are also included as an Appendix for convenient reference. But there is no intention here to suggest that Needham's views are more authoritative than other interpretations. By contrast, Kabbalah as a subject for scholarly research is due to the monumental work of Gershom Scholem. He is thus the major source for the discussion of the Tree,[18] though this essay also draws on the work of Idel and other Kabbalah scholars. Relying on these prominent sources must suffice since, as both Idel and Abrams[19]

There is a distinct similarity in the worldliness and moral focus of Confucianism (less salient in both Taoism and Buddhism) and rabbinic Judaism. Worldliness and moral focus was reinvigorated in both traditions by excursions into cosmology/theosophy and esoteric spirituality.

17. Needham, *Science and Civilization in China*, 460–64. For other translations, see, e.g., Bruce, *Chu Hsi [Zhu Xi] and His Masters*, 128–33; Fung, *A History of Chinese Philosophy*, 435–38 (Zhou Dunyi's commentary).

18. Other Scholem works that have been consulted are *Major Trends in Jewish Mysticism*; *On the Kabbalah and Its Symbolism*; *Origins of the Kabbalah*.

19. Idel (*Kabbalah: New Perspectives*, 136) writes: "there is as yet no comprehensive study of the history of the Kabbalistic doctrines of the Sefirot." Daniel Abrams concurs ("New Study Tools from the Kabbalists of Today").

note, there is yet no definitive treatment of the history of the doctrine of the *Sefirot* and their use in Kabbalistic structures.

MEANING AND SEQUENCE

The sequence of components in the Diagram is:

- (I) *Taiji* ("the Supreme Pole");
- (II) the Two Forces, Yang and Yin;
- (III) the Five Agents;
- (IV) *Qian* and *Kun* "(*Ch'ien* and *K'un* in Figure 5.1);
- (V) the myriad things.

The connection between the Forces and the Agents is not itself a separate element; nor is the small circle at the bottom of the Agents. The sequence in the Tree (the *Sefirot* are numbered from right to left) is:

(1) *Keter*, Crown

(3) *Binah*, Understanding, Intelligence

(2) *Hokhmah*, Wisdom

(5) *Din*, Judgment, Law, Rigor

(4) *Hesed*, Love, Mercy

(6) *Tifereth*, Beauty, Splendor

(8) *Hod*, Majesty

(7) *Netsah*, Eternity, Endurance

(9) *Yesod*, Foundation

(10) *Malkhuth*, Kingdom

Sometimes a supplementary *Sefira*, *Da'at*, Knowledge, was interposed between *Hokhmah-Binah* and *Hesed-Din*, but this was not numbered among the canonical *Sefirot*.[20]

---

20. Scholem, *Kabbalah*, 107.

The First Three Elements

Both symbols begin at the top with a neutral element representing the highest reality: *Taiji*, the Supreme Pole in the Diagram and *Keter*, Crown, in the Tree. Yet the identity of this first element is not free of ambiguity. Zhou Dunyi and Zhu Xi both note that, "The Supreme Pole is essentially (identical with) that which has no Pole." There are two concepts here: the Supreme Pole, *Taiji*, from the Confucian (and Daoist) classic, the *Yijing*, and "that which has no Pole," the "Ultimateless," *Wuji*, from the *Dao Dejing*.[21] The identity of these "positive" and "negative" ("full" and "empty") concepts is asserted in the commentaries, but these concepts were not completely synonymous. As Henderson points out,[22] the identification of *Taiji* and *Wuji* is a syncretic statement uniting notions from different Chinese traditions.

A parallel union of positive and negative concepts existed in the Kabbalah in the relationship between *Keter* and—not included in the symbol—*Ein-Sof*.[23] In some Kabbalist writings, *Ein-Sof*, "that which has no end," is more fundamental than *Keter* and beyond description. In other writings, *Keter* is the external aspect of *Ein-Sof*, indicating a closer relationship. *Keter* is also referred to as *Ayin*, "nothingness," a negative concept like *Ein-Sof*, whose polar opposite is *Yesh*, existence, literally "there is."[24] There is a relationship between that which is manifested—*Keter*—and that which is unmanifested—*Ein-Sof* or *Ayin*: *Yesh* arises from *Ayin*, Being from Nothingness. In both traditions, beyond what can be stated as the highest is that which has no name, no end, no pole. Both traditions wrestled with the problem of whether the unmanifested is prior to and distinct from the manifested, or whether the two are in some sense equivalent. Neither the solution of difference nor the solution of identity was completely satisfactory, and so different positions inevitably arose on this matter. It is not being asserted here that *Wuji* is identical with *Ein-Sof* or *Ayin* (although *Wuji* means "no extreme," quite close to *Ein-Sof*, which means "no end"). Virtually every mystical tradition has some notion of Nothingness, as doctrine and as meditative or mystical experience. While notions of Nothingness in different cultures

---

21. Needham, *Science and Civilization in China*, 464.
22. Henderson, *Development and Decline*.
23. Scholem, *Kabbalah*, 88–92; Tishby, *Wisdom of the Zohar*, 235 ff.
24. Matt, "Concept of Nothingness," 67–108.

are not the same, it is equally implausible to believe they are completely different. Both Neo-Confucians and Kabbalists faced the question of the relationship between Nothingness and Plenitude. Corresponding terms do not mean the same thing—*Ein Sof* and *Keter* are theistic concepts but *Wuji* and *Taiji* are not[25]—but the relation between *Wuji* and *Taiji* and the relation between *Ein-Sof* and *Keter* are similar.

In both symbols, the first element gives rise to a dyad representing the fundamental polarity that emanates from the fundamental unity: for the Diagram, the Two Forces, Yang and Yin; for the Tree, *Hokhmah*, Wisdom, and *Binah*, Understanding or Intelligence. In this dyad, the male element is first and the female element second. Zhou Dunyi writes, "The Supreme Pole moves and produces the Yang. When the movement has reached its limit, rest (ensues). Resting, the Supreme Pole produces the Yin." Correspondingly, Wisdom and Understanding are second and third in the canonical order of the *Sefirot*. But one should not make too much of this ordering. The placement of Yang and Yin and Wisdom and Understanding implies symmetry for the two elements; for the Diagram, this symmetry also inheres in the fact that Yang generates Yin and Yin generates Yang. There is a tension here between asserting symmetry and breaking symmetry (sequencing the elements); both are required. The first three elements in each structure constitute a primary triad from which the rest of the symbol follows. In Daoist thought, the union in the Dao of Yin and Yang was an explicit triad, and this was incorporated into Neo-Confucian philosophy. In the Tree, this triad is also recognized as an explicit unit and the generative source from which creation proceeds.[26] Both triads represent the differentiation of unity into duality with a resulting symbolism of one, two, and three, rooted in an ineffable zero, empty yet also full.

The Yin character of Understanding was prominent in Kabbalist thought. While the tenth *Sefirah* of *Malkhuth*, Kingdom, represented the *Shekhinah*, the "Divine Presence" and female aspect of God, there was

---

25. *Wuji* and *Taiji* might qualify as "philosophically theistic": Wang observes (Zhou Dunyi's Diagram of the Supreme Ultimate Explained, 318) that Fung (*History of Chinese Philosophy*, 537), commenting on Zhu Xi's interpretation of Zhou Dunyi, says: "Spoken of in this way the Supreme Ultimate is very much like what Plato called the Idea of the Good, or what Aristotle called God." But Wang insists that, "the differences are equally fundamental. *Wuji/Taiji* is emphatically nontheistic, for it cannot be understood as God in any way that might confuse it with the specific teachings of 'classical theism.'"

26. Scholem, *Kabbalah*, 108; Scholem, *On the Kabbalah and Its Symbolism*, 103.

a doctrine of a higher and a lower *Shekhinah*, of which the higher was Understanding and the lower was Kingdom. Scholem writes, "As the upper *Shekhinah* of the *Sefirah* of *Binah*, [the principle of] femininity is the full expression of ceaseless creative power—it is receptive, to be sure, but is spontaneously and incessantly transformed into an element that gives birth, as the stream of eternally flowing divine life enters into it."[27]

In both symbols, the first three elements encompass the distinction between form and substance, although they do so in different ways. Zhu Xi linked the Supreme Pole itself (circle I) with *Li*, principle, whose original meaning was "order" or "pattern," sometimes equated with Aristotelian "form."[28] *Li* is interpreted by Needham in scientific terms as "organization," in contemporary scientific language, "information,"[29] and Yin and Yang (circle II) with *Qi*, interpreted by Needham as "matter-energy," which accords with the inherent generativity of the Two Forces; *Li* and *Qi* are inherently linked, as information is always associated with matter-energy. In the Tree, however, the form-substance distinction is *not* in *Keter* vs. *Hokhmah* and *Binah*, but rather in *Hokhmah* vs. *Binah*. Scholem notes, "This conception formulated by Plato in the *Timaeus*, where *hyle* [matter] is called mother and form [*morph*] is called father, corresponds to symbolism commonly used among the Kabbalists for *Hokhmah* and *Binah*."[30]

The Five Agents and the Central Sefirot

The middle portion of the Diagram consists of the Five Agents, Fire, Water, Earth, Wood, and Metal.[31] Zhou Dunyi writes, "The Yang is transformed (by) reacting with the Yin and so Water, Fire, Wood, Metal, and Earth are produced." For Zhu Xi, the order is Water-Wood-Fire-Earth-Metal. The Five Agents are not material entities but rather are processes that are fire-*like*, water-*like*, etc. In modern terms, they are functional and

---

27. Scholem, *On the Mystical Shape of the Godhead*, 174.

28. Patt-Shamir, *To Broaden*, 232.

29. Needham, *Science and Civilization in China*, 472 ff. In Needham's interpretation of *Li* and *Qi* as organization (information) and matter-energy, one can also see an echo of the Hindu gunas: Sattva (intelligence) is *Li*, and Rajas (energy) and Tamas (material inertia) are joined together as the Yang and Yin of *Qi*.

30. Scholem, *Origins of the Kabbalah*, 1987, 428n.

31. Needham's translation of "Five Elements" is replaced here by the more common "Five Agents."

abstract and reflect a "stuff-free" systems-theoretic viewpoint. Just as systems theories focus on modes of organization and process for which the materiality of the phenomena described is not important,[32] the names of the Agents are concrete illustrations that are not intended literally. (The same can be said of "four elements" ideas in Greek and medieval thought.) Agents are categorized as major and minor Yang (Fire and Wood), major and minor Yin (Water and Metal), and neutral (Earth). They are ordered by a number of different sequences, and the main ones are given in Table 5.1. In graph-theoretic language, these sequences are 'directed graphs' ('digraphs') that are either cyclic or acyclic.

TABLE 5.1. Enumeration Orders of the Five Agents (Needham[33])

The repetition of Wood in the 2nd and 3rd order indicates the cyclicity of these two orders.

The Cosmogenic Order	Water-Fire-Wood-Metal-Earth
The Mutual Production Order	Wood-Fire-Earth-Metal-Water-(Wood-...)
The Mutual Conquest Order	Wood-Metal-Fire-Water-Earth-(Wood-...)
The "Modern" Order	Metal-Wood-Water-Fire-Earth

Zhou Dunyi's commentary on the Diagram uses the acyclic Cosmogenic Order, while Zhu Xi's commentary uses the cyclic Mutual Production Order, starting with Water. In the Diagram as shown in Figure 5.1, Earth is directly connected to both Fire and Metal, and Water and Wood are also directly connected, which points to the Mutual Production Order. Needham notes that the relations of "production" and "conquest" are very close to modern scientific ideas; indeed these ideas are standard in causal (directed graph) analysis.[34] Needham's view

---

32. The idea of a "stuff-free" science is from Mario Bunge's *Method, Model and Matter*, ch. 2 (Testability Today), ch. 8 (Is Scientific Metaphysics Possible?).

33. Needham, *Science and Civilization in China*, 253 ff. The Cosmogenic Order is the "evolutionary order in which the elements [agents] were supposed to come into being." In the Mutual Production Order, Fire is produced (increased) by Wood, Earth by Fire, etc. In the Mutual Conquest Order, Wood is "conquered by" Metal, Metal by Fire, etc. Needham says that the Modern order is obscure and primarily of popular and not philosophical significance.

34. Given some A→B relation, interpreted either as (i) $dB/dt = k A$ or as (ii) $B = k A$, for some constant k, the relation is one of 'production' when k is positive and one of

of early Chinese thought as proto-scientific, and—from the perspective of this paper—as a non-mathematical precursor of systems theory, is especially appropriate to the doctrine of the Five Agents.[35]

The middle portion of the Tree are the five *Sefirot*: *Hesed*, Benevolence (Love, Mercy; or *Gedulah*, Greatness); *Din*, Judgment (Law, Rigor; or *Gevurah*, Power);[36] *Tiferet*, Beauty (Splendor; or *Rahamim*, Compassion); *Netsah*, Eternity; and *Hod*, Glory (Majesty). Benevolence and Eternity are primary and secondary male *Sefirot*, Judgment and Glory are primary and secondary female *Sefirot*, and Beauty (6) is neutral. Here a major difference exists between the symbols: the substructure of the Five Agents is plain in the Diagram, but an explicit pentad of Benevolence to Glory does not appear in the Tree or in Kabbalist literature. While the symbolism of five was salient in Chinese philosophy, it was largely absent in Jewish thought,[37] although it existed in occult Kabbalah.[38]

'conquest' when k is negative. For an odd number of relations of type (i), cycles consisting only of relations of production or only of relations of conquest (the second and third orders of Table 5.1) are examples of positive and negative feedback loops, respectively. Complex systems encompass loops of both types, and their analysis normally requires knowing the magnitudes of the k's for all the individual relations. In special cases, however, knowing only the signs of the k's—such systems are called "signed digraphs"—suffices to determine the overall dynamic behavior; see Richard Levins, "The Qualitative Analysis of Partially Specified Systems."

35. Needham's view applies also to ideas and diagrams associated with the *Yijing*; see Ryan, "Leibniz's Binary System and Shao Yong's *Yijing*."

36. *Din* is chosen here although *Gevurah* is more common for this *Sefirah*, because Figure 5.1 uses *Din*, and because the meaning of *Din* is clearer.

37. Needham, *Science and Civilization in China*, 297. In Kabbalistic ideas about hierarchical components of the soul (*Nefesh*, *Ruach*, and *Neshamah*), one can find *Ruach* sometimes identified with the six *Sefirot*, Benevolence through Foundation and sometimes simply with Beauty. According to Tishby (120 ff.), this tripartite conception is the prevailing view of the soul in the *Zohar*, the central book of the Kabbalah. Most commonly, *Nefesh* is the lowest component of the soul, *Neshamah* the highest, and *Ruach* is intermediate between the two. The traditional assignments were *Neshamah* to Understanding, *Ruach* to Beauty or to Benevolence through Foundation, and *Nefesh* to Kingdom, but Tishby notes that the Kabbalist literature is not at all consistent in the correlations of *Sefirot* to these components of soul. Sometimes other components (*Chiah* and *Yechidah*) were added, usually as still higher levels of the soul (Scholem, 1974, 157). Roughly, then, *Ruach* is associated with the middle portion of the Tree, approximately analogous to Five Agents in the Diagram, but the correspondence is far from exact. There do not appear to be pentadic groupings parallel to the Five Agents in Kabbalistic correlations of planets with the *Sefirot*, or in the doctrine of the four "worlds" (*Atziluth*, *Briah*, *Yetsirah*, *Assiah*).

38. Occult Kabbalah had a developed symbolism of five, and Regardie associated *Ruach* with Benevolence through Glory (Regardie, *Garden of Pomegranates*). Regardie

If one aligns major and minor Yang Agents with primary and secondary Male *Sefirot*, and major and minor Yin Agents with primary and secondary Female *Sefirot*, one obtains the correspondences of Fire-Benevolence, Water-Judgment, Earth-Beauty, Wood-Eternity, and Metal-Glory, as shown in Table 5.2. The sequence of Agents, following the canonical order of the *Sefirot*, is Fire-Water-Earth-Wood-Metal, i.e., the Mutual Conquest Order starting with Fire.

TABLE 5.2. The Five Agents and *Sefirot* 4–8

Agents		Sefirot	
Yin	Yang	Female	Male
Water	Fire	Judgment (5)	Benevolence (4)
	Earth	Beauty (6)	
Metal	Wood	Glory (8)	Eternity (7)

A more interesting parallelism, however, aligns the central *Sefirot* with the Chinese pentad of Five Virtues, as shown in Table 5.3. These are the primary Yang and Yin virtues of (a) *Ren*, Benevolence (Humanity, Love) and (b) *Yi*, Righteousness (Rightness), (c) the neutral virtue of *Xin*, Sincerity (Honesty, Good Faith, Trustworthiness), and the secondary Yang and Yin virtues of (d) *Li*, Reverence (Propriety; not the same as but related to *Li*, Principle) and (e) *Zhi*, Wisdom; these are associated with Wood, Metal, Fire, Water, and Earth, respectively. This mirror-reflects the Five Agents, correlating primary and secondary *Sefirot* with major and minor Virtues instead of major and minor Agents.[39]

---

claims this conception of *Ruach* is "essentially derived" from Rabbi Azriel of Gerona, a pupil of Isaac the Blind, but this claim is not consistent with Tishby's (*Wisdom of the Zohar*, 32) assertion that the Rabbi Azriel's five parts of the soul "originated from the first five *Sefirot*."

39. Although *Ren* and *Yi* are the major Virtues, for some reason they are assigned to the minor Yang and Yin elements, *Wood* and *Metal*. The sequence of Agents obtained in this way, following the order of *Sefirot*, is Wood-Metal-Earth-Fire-Water, which is the Modern Order taken as cyclic (though Needham gives this order as acyclic) and in reverse, starting with Wood. This is plainly not a canonical order. Still, aligning major and minor Virtues with primary and secondary *Sefirot* does still yield a plausible correlation.

TABLE 5.3. The Five Virtues and *Sefirot* 4–8

Virtues		Sefirot	
Yin	Yang	Female	Male
Yi	Ren	Din	Hesed
Righteousness (Metal)	Benevolence (Wood)	Judgment	Benevolence
	Xin		Tiferet
	Sincerity (Earth)		Beauty
Zhi	Li	Hod	Netsah
Wisdom (Water)	Reverence (Fire)	Glory	Eternity

The pentad of Virtues was central to the transformation of the Daoist precursor of the Diagram to its Neo-Confucian form. In the earlier Daoist version, the Agents referred to aspects of meditation, but for Zhou Dunyi—and Zhu Xi agrees[40]—their primary relevance was to the Virtues and the achieving of sagehood: "The sages ordered their lives by the Correct, by Love and Righteousness. They adopted ataraxy as their dominant attitude, and set up the highest standards for mankind. Thus it was that the 'virtue of the sages was in harmony with that of heaven and earth'... The good fortune of the noble man lies in cultivating these virtues; the bad fortune of the ignoble man lies in proceeding contrary to them."

The Diagram was a metaphysical basis for ethics.[41] As Zhou Dunyi writes, it was the harmonious development of the Virtues (component III) which provided the basis for the distinction between good and evil (circle IV). Human conduct remained the central concern of the Neo-Confucians, however much they were influenced by the spiritual focus

---

40. For Zhu Xi, see Chiu Hansheng, "Zhu Xi's Doctrine of Principle," in Wing-tsit Chan, *Chu Hsi and Neo-Confucianism*, 129–35.

41. Teng Aimin, "Chu Hsi's [Zhu Xi's] Theory of the Great Ultimate," in Wing-tsit Chan, *Chu Hsi and Neo-Confucianism*, 110. Welch expressed this idea directly: "This Neo-Confucianism... developed because Confucius had never formulated a metaphysics and the lack of it put his later followers at a disadvantage in their rivalry with the complete philosophical systems of Taoism and Buddhism" (Welch, *Taoism*, 158.) Welch also quotes Fung as saying that the Neo-Confucians were "more Taoistic than the Taoists and more Buddhistic than the Buddhists."

of Buddhism and Taoism.⁴² While meditation ("quiet-sitting") provided a means of self-cultivation, it was not viewed as an end in itself. Shu-Hsien Liu notes that "the Buddhists' ultimate commitment is . . . *Shunya* or Emptiness," but the "ultimate commitment for the Confucianists [remained] *Ren* (Humanity)."⁴³

In this pentad of Virtues, *Ren* and *Yi* is the principal dyad, the first Yang and the second Yin. Benevolence is primary, and all other virtues, especially Righteousness, flow from it. So too in the Tree, *Hesed* (Benevolence) is prior to and the source of *Din* (Judgment), the first being masculine, the second feminine. Fung notes that Righteousness was "the goodness that comes from hardness" and included "decisiveness, strictness, firmness, determination, and steadfastness,"⁴⁴ which are also the qualities of *Din*. Also, the predominance of *Ren* and *Yi* over the other three Virtues matches the predominance of *Hesed* and *Din* over the following three *Sefirot*. But it is not being asserted here that *Ren* and *Hesed* are identical, despite the appropriateness of the translation 'benevolence' for both, or that *Yi* and *Din* are identical. *Ren* is rooted in the different human relationships (father–son, ruler–subject, etc.) whose specific obligations are emphasized in Confucianism, but understood as "benevolence" *Ren* transcends these relationships. *Ren* was the subject of extensive scholarly discourse in Confucianism, and the concept of *Hesed* was similarly complex. Still, with respect to the male-female polarity, *Ren* and *Yi* clearly parallel *Hesed* and *Din*. What is especially interesting in this parallelism is that, contrary to popular Western gender correlations, both Jewish and Chinese medieval philosophy assigned mercy to the

---

42. The conceptualization of the Virtues was influenced by these "more spiritual" traditions. For example, Julia Ching notes that Zhu Xi speaks of "abiding in Reverence, defining it in terms of single-mindedness and freedom from distraction and comparing it to the Buddhist practice of mindful alertness" (Ching, "Chu Hsi," 280). Ching goes on to compare the practice of Reverence to the "recollection" of Western Christian spirituality.

43. Liu, "Orthodoxy in Chu Hsi's Philosophy" in Wing-tsit Chan, *Chu Hsi and Neo-Confucianism*, 441.

44. Fung, *A History of Chinese Philosophy*, 447.

masculine and severity to the feminine.[45] Both Jewish and Chinese thinkers also regarded imbalance within these dyads as a source of evil.[46]

One might see parallels between Reverence and Eternity (Zhu Xi reinterpreted Reverence as mindfulness, collectedness, a kind of dwelling in eternity) and between Wisdom and Glory (both of which give content to this dwelling). Sincerity and Beauty, neutral in polarity, center and "give reality" and dynamism to adjacent elements. But these correlations seem less compelling than the *Ren-Hesed* and *Yi-Din* correlations.

Aligning the Chinese pentad of Virtues with the central *Sefirot* according to Table 5.3 has a consequence that is intriguing, though it would be hard to argue that this is not mere coincidence. At the bottom of the Five Agents in the Diagram, there is a small circle that is not an element in its own right, but about which Zhu Xi writes, "The small circle below, connected by the four lines with the Five Agents above, indicates that which has no Pole, in which all are mysteriously unified." If Wood and Metal are placed at the top of the Five Agents as displayed in Table 5.3, the small circle is then above them, precisely at the site of the

---

45. The matter is not as simple as this. Fung notes that Righteousness was "the goodness that comes from hardness," and this is supported by Zhou Dunyi's comment, "Therefore it is said, 'In representing the Tao of Heaven one uses the terms Yin and Yang, and in representing the Tao of Earth one uses the terms Soft and Hard; while in representing the Tao of Man, one uses the terms Love and Righteousness." Yet the Virtues of Benevolence and Righteousness are Yang and Yin, respectively, not the reverse, which these quotes seem to imply.

46. Scholem, *On the Mystical Shape of the Godhead*, 1991, chapter 2: Good and Evil in the Kabbalah, and Fung, *A History of Chinese Philosophy*, 446–47, discussing Zhou Dunyi's commentary. Virtues—more precisely, their absence—is about "moral evil," rather than a more general "metaphysical evil"—this distinction being one commonly made by Western philosophers—but metaphysical evil was also of concern to both Kabbalists and Neo-Confucians. Indeed, one might say that in both traditions, moral good and evil *are* metaphysical. In both traditions, there is another account of the origin of evil that does not attribute it to an imbalance between Benevolence and Righteousness (Judgment) but instead locates it at a higher level. According to Fung (*A History of Chinese Philosophy*, 552–56) Zhu Xi's views on this resemble Plato's notion that imperfection arises from the material instantiation of the Ideas (Forms). What corresponds to the Ideas is Principle (*Li*), which is *Taiji*, where according to Zhu Xi perfection reigns. What adds materiality—and hence imperfection—to all manifestations are the Two Forces. In this view, it is in the transition from level I to level II that evil in introduced into the cosmos. The top portion of the Tree is also implicated in metaphysical evil. In Nahmanides' early form of the Lurianic *tsimtsum*, the contraction of God that is necessary for Creation, the ultimate source of metaphysical evil, is located in a disruption caused by *tsimtsum* , not in *Ein-Sof* but in Crown in its origination of Understanding (Scholem, *Origins of the Kabbalah*, 449).

"supplementary" *Sefirah* of *Daʾat*, Knowledge—not numbered among the canonical *Sefirot*[47] and not shown in Figure 5.1—that is sometimes interposed between Wisdom-Understanding and Benevolence-Judgment.

The Last Two Elements

The last two elements of both symbols are neutral in gender: in the Diagram, circle IV, *Qian* and *Kun*, and circle V, the myriad things; in the Tree, *Yesod*, Foundation, and *Malkhuth*, Kingdom. In both, the next to last element is the sexual generative power and the funnel through which all elements above merge and flow into the final element. The last element is the multiplicity of all things which results from this influx via the union of sexual powers.

The sexually generative character of the last two circles of the Diagram is asserted by both Zhou Dunyi and Zhu Xi. "The Two Qi (of maleness and femaleness), reacting with and influencing each other change and bring the myriad things into being. Generation follows generation, and there is no end to their changes and transformations." (Zhou Dunyi)

"The fourth figure represents (the operations of the Qi of Yin and Yang exhibited in) the principles of (heavenly) maleness and of (earthly) femaleness which pervade the universe ... The fifth figure represents the birth and transformation of the myriad things in their sensible forms, each of which has its own nature." (Zhu Xi)

Qian and *Kun*, the male and female aspects of circle IV, are the primary Yang and Yin trigrams and hexagrams in the *Yijing*; they consist exclusively of Yang and Yin lines, respectively.[48] This circle thus links the Diagram to this Confucian classic which Zhou Dunyi says "is the most perfect." While Yin and Yang are not generally sexual, in circle IV they are. Needham states that Zhou Dunyi's commentary on circle IV is "undoubtedly chemical, cf., the sexual symbolism of the alchemists."[49] In the Daoist antecedent of the Diagram, used to guide meditation, the commentary on circle IV is explicitly alchemical; Zhou Dunyi retained this association. About the Tree, Scholem writes: "The ninth Sefirah, *Yesod*, is the male potency, described with clearly phallic symbolism, the 'founda-

---

47. Scholem, *Kabbalah*, 107.

48. Fung, *A History of Chinese Philosophy*, 454–56.

49. Needham, *Science and Civilization in China*, 461. Fung (*History of Chinese Philosophy*, 441) concurs.

tion' of all life, which guarantees and consummates the hieros gamos, the holy union of male and female powers."[50]

Foundation has a masculine character in relation to Kingdom, but it is not exclusively masculine, as its placement on the central column attests. The phallic symbolism comes from using the male figure to associate *Sefirot* with bodily parts, but genital symbolism is really intended. Scholem notes, "The ninth Sefirah, *Yesod*, 'the foundation,' is correlated with the male and female sex organs . . . out of which all the higher *Sefirot*—welded together in the image of the King—flow in to the *Shekhinah*, [and] is interpreted as the procreative life force dynamically active in the universe."[51]

Sexual rites and meditations were associated with Foundation. Scholem quotes a Friday evening hymn of Isaac Luria, the great Safed Kabbalist, which speaks of the union of husband and wife and makes this quite explicit.[52] The argument here is not that there was a sexual alchemy within Kabbalah[53] but that the sexual symbolism of Foundation resembles the sexual aspect of Chinese alchemy.

A moral dimension of circle IV augments its sexual aspect. Zhou Dunyi writes, "It is man alone, however, who receives the finest (substance) and is the most spiritual of beings. After his (bodily) form has been produced, his spirit develops consciousness; (when) his five agents are stimulated and move, (there develops the) distinction between good and evil, and the myriad phenomena of conduct appear."

The distinction between good and evil is circle IV; the "myriad phenomena of conduct" which flow from this distinction is circle V. *Qian*, the Yang aspect of circle IV, is associated with sincerity, which for Zhou Dunyi is the basis, the beginning, of sagehood,[54] the sage being the high-

---

50. Scholem, *On the Kabbalah and Its Symbolism*, 104.

51. ibid, 143 and 227.

52. Ibid., 143.

53. While Patai has documented evidence of Jewish involvement in alchemy since at least the Hellenistic era (100 BCE to 100 CE), he does not indicate that any sexual aspect was prominent in Jewish alchemy either in this period or much later, when alchemy was influenced by Kabbalah (Patai, *Jewish Alchemists*). In the later alchemical use of Kabbalah, Foundation does not appear to have been singled out for special attention.

54. Patt-Shamir, *To Broaden*, 174, quotes Zhou Dunyi as saying in his Book of Comprehensiveness (*Tongshu*) that "sincerity is the foundation of the sage." The sincerity being spoken of here is *cheng*, not *xin*, correlated with Earth, which Patt-Shamir translates instead as trustworthiness.

est human moral ideal in Confucianism. Similarly Foundation is also called *Zaddik*, "the righteous one," the zaddik being the highest moral ideal of Judaism: Righteousness is the foundation of the world,[55] and is associated with moral distinctions and harmonious equilibrium, with setting things in their proper places. (The Righteousness of the *Sefirah* Judgment is a more general concept, meaning also Rigor and Power; the Righteousness of Foundation refers to specific behavior.) There is also a moral connection to the sexual aspect of Foundation. This *Sefirah* was associated with the biblical figure of Joseph, who resisted sexual temptation.

The symbolism of the last element is also similar. Circle V, the "myriad things"[56] is the multiplicity finally engendered by the Supreme Pole.[57] Although this circle is not considered Yin by either Zhou Dunyi or Zhu Xi, in the Daoist precursor of the Diagram it is called the "Doorway of the Mysterious Female" or "The Gate of the Dark Femininity."[58] Circle V corresponds to Kingdom, which unites the *Sefirot* and represents the attribute of God linked most closely with the Material World. Kingdom is distinctively female, corresponding to the lower *Shekhinah*, the female aspect of God, the divine immanence within the multiplicity of existence. It is "in everything" (*ba-kol*), the "form that embraces all forms" and renders to each form its specific individuality.[59] Plurality is also reflected in the interpretation of this last *Sefirah* as representing "*Knesset Israel*," the mystical archetype of the community of Israel.[60]

The last element is farthest from the first, and is a terminus, yet like the other elements, it remains connected to its source. The words of the

---

55. Scholem, *Kabbalah*, chapter 3: Tsaddik: The Righteous One.

56. Also translated as the "ten thousand things" (Fung, *History of Chinese Philosophy*, 445), a concept that dates at least back to the *Dao Dejing*, and used in Chinese thought to indicate the multiplicity of existence. There is a possible Jewish parallel. Joseph Dan, in his *The Ancient Jewish Mysticism* (p. 74) writes, "Ancient Hebrew, as modern-day Hebrew, does not have a word for any number larger than 10,000. Today, when we wish to discuss astronomical distances or deal with the state budget, we are forced to use Latin terms: million, billion, etc. The Hebrew horizon did not extend beyond 10,000."

57. This multiplicity is different from the multiplicity generated by the binary exponentiation of the *Yijing*. The Diagram treats this latter multiplicity as a unity by its referring to the *Yijing* with the simple circle IV.

58. Fung, *History of Chinese Philosophy*, 441; Chang (*Creativity and Taoism*, 166). The concept comes from Laozi.

59. Scholem, *On the Mystical Shape of the Godhead*, 171, 179.

60. Scholem, *Origins of the Kabbalah*, 167–69.

Sefer Yetsirah 1:7, "Ten *Sefirot* of Nothingness: Their end is imbedded in their beginning and their beginning in their end"[61] resembles Zhu Xi's commentary on circle V, "But (as indicated again by reproduction of the original circle) all the myriad things go back to the one Supreme Pole." (The point is weakened by Zhu Xi saying the same thing about circle IV, but he means that all the elements of the Diagram are united in their source, as was also held by the Kabbalists about the *Sefirot*.) Circularity in the Diagram is also suggested by its mirror symmetry: circle V mirrors circle I and circle IV mirrors circle II (Yang and Yin being inside circle II makes this possible). In the Tree, circularity is suggested by Kingdom being related in meaning to the first *Sefirah*, Crown. Kingdom is also called *Atarah*, another word for crown.[62] The Tree, however, is visually less symmetric because Wisdom and Understanding are structurally separate, unlike Yang and Yin in circle II of the Diagram.

OVERALL ARCHITECTURE

If one steps back from the elements and their relationships and looks at the overall architecture of the symbols, one sees that their global structures, the hierarchical sequence of levels and the spatial arrangement of male, female, and neutral elements, are very similar. The vertical hierarchy in each symbol articulates levels of differentiation from the primal unity to the multiplicity of existence, but this progression does not imply a simple directionality that privileges the higher elements. Like the tension between symmetry vs. asymmetry (e.g., sequence, gender polarity) for elements at the same level, there is tension between hierarchy (directionality) vs. non-hierarchy in the relations between levels. Although levels reflect a progression, the circularity of the symbols counters directionality. Moreover, Zhu Xi insists that "the Supreme Pole ... should be regarded neither as separate from, nor as identical with, the Two Forces ... The Five Agents all come from the Yin and Yang (Forces). The five different things (fit into) the two realities without the slightest excess or deficiency. And the Yin and the Yang (go back to) the Supreme Pole (perfectly), neither one of them being more or less elaborate than the other, nor more or less fundamental than the other." (Yet Zhu Xi

---

61. Aryeh Kaplan (*Sefer Yeszirah*, 57) notes that "beginning" refers to Crown and "end" to Kingdom, and explicitly offers a circular visualization of their connection.

62. Scholem, *On the Possibility of Jewish Mysticism*, 143.

affirms that the Five Agents and the myriad things all have their "specific natures," which is not said by him about *Taiji* or Yin-Yang, suggesting a difference that still distinguishes the elements.) While the Kabbalists did not stress the equality of all parts of the Tree, homogeneity is suggested in the multiple polar dyads of the neutral column: Crown-Kingdom, Beauty-Kingdom, and Foundation-Kingdom. (There are no vertical polar dyads in the Diagram.) Crown is echoed in Beauty, Foundation, and Kingdom.

The elements of both symbols can be assigned to male, female, and neutral vertical columns. Classifying entities as male, female, or neutral was a ubiquitous feature of traditional thought, and Needham noted the tendency in Kabbalah to arrange lists of pairs in a manner similar to the Chinese Yin-Yang categories.[63] In the Diagram, the columns are not explicit, but the principle is clear. Yang, associated with expansion,[64] encompasses Fire (major Yang) and Wood (minor Yang). Yin, associated with concentration, encompasses Water (major Yin) and Metal (minor Yin). The central neutral column includes circles I, IV, and V, and Earth, which is a synthesis of Yin and Yang. For the Tree (left-right assignments are reversed relative to the Diagram), the columns are quite explicit: the right column includes Wisdom, Benevolence, and Eternity, the left column Understanding, Judgment, and Glory, and the central column, includes Crown, Beauty, Foundation, and Kingdom. The right and left columns represent male and "expansive" versus female and "concentrative" attributes of God.[65] The central column is neutral but includes the vertical gender polarities mentioned above.

One can alternatively see the structures as consisting of horizontal male-female dyads[66] often elaborated by the introduction of a third element representing either[67]

---

63. Needham, *Science and Civilization in China*, 297.
64. Ibid., 471.
65. Frank, *The Kabbalah*, 106.
66. Needham, *Science and Civilization in China*, 297.
67. These two types of triad are discussed by René Guénon (*Great Triad*). The differentiating triad (Figure 5.2a) is a transition from the monad to the dyad. The integrating triad (Figure 5.2b) resembles Bennett's (*Dramatic Universe*) "evolutionary" triad of creation, in which an active element interacts with a passive one to yield a neutral result.

a. the origin of the dyad, i.e., the (higher) unity of which they are (lower) parts; this manifests *differentiation* (Figure 5.2a); or

b. a (lower) synthesis which reconciles their (higher) opposition; this manifests *integration* (Figure 5.2b).

FIGURE 5.2. Differentiating (a) and Integrating (b) Triads

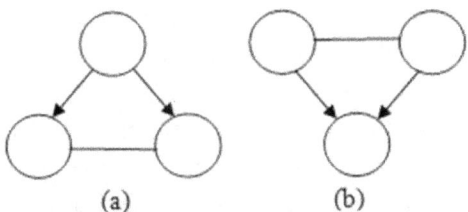

Differentiation is illustrated in the Diagram by the relation between *Taiji* and the dyad of Yang and Yin, and in the Tree by the relation of Crown with Wisdom and Understanding. Integration is illustrated in the Tree by the triads of Benevolence-Judgment-Beauty and Eternity-Glory-Foundation. Integrating triads in the Diagram are less apparent; Earth might be considered a synthesis of major and minor Yin and Yang Agents, but this synthesis is not triadic, and circle IV derives from all the Five Agents rather than from any single Yin-Yang dyad. However, there is a triad implicit in the relation between the two aspects of circle IV with circle V: *Qian* (Heaven, primary Yang) and *Kun* (Earth, primary Yin) unite to generate the "myriad things," but this triad is not explicit since circle IV is visually a monad, not a dyad, like circle II.

Symbolic triads were widely prevalent in both East and West, so it is not surprising to see such triadic schemes in these Chinese and Jewish symbols. What is remarkable is that the union of hierarchical and polar organizing principles produces an identical spatial distribution of elements: proceeding downward, both symbols begin with a neutral element, which splits into a male-female dyad, from which are derived a dyad, a neutral element, and another dyad, after which the symbol is completed by two neutral elements.

The Diagram and Tree have the same or nearly the same number of elements. The Tree is explicitly constructed from the ten *Sefirot*. The number ten had great symbolic resonance in Jewish thought, and the *Sefer Yetsirah* explicitly insisted upon this precise number: "Ten and not nine; ten and not eleven."[68] The Diagram is also composed of ten elements if one counts Yang and Yin, the parts of circle II, as two elements, which is suggested by the Two Forces being visually distinct, and if one counts circle IV as one element, since two-foldedness is not visually indicated. But it is unnecessary to insist that the Chinese structure has precisely ten elements. It is the similarity of this structure to the Tree, not its number of elements, which is interesting. While the symbolism of two and three is found in both traditions, the symbolism of ten is a Western one, being present in Jewish, Pythagorean, Gnostic, and early Christian writings, and is not indigenous to Chinese thought. (It was, however, prominent in Indian thought which passed into China through Buddhism.)

The Tree was sometimes also conceptualized as a triad Crown-Wisdom-Understanding, followed by a heptad of the remaining seven "*Sefirot* of Construction," or as three triads (Crown-Wisdom-Understanding pointing up, and Benevolence-Judgment-Beauty and Eternity-Glory-Foundation pointing down) leading to and summarized in Kingdom,[69] or as a monad (Crown), followed by an octad (Wisdom to Foundation), completed by a monad (Kingdom).[70] Other spatial configurations appear in the history of the symbol,[71] and there are also different representations of the channels connecting the *Sefirot*.[72]

68. Scholem, *Origins of the Kabbalah*, 144.
69. Scholem, *Kabbalah*, 107–9.
70. Idel, *Kabbalah: New Perspectives*, 55.
71. See footnote 6. The possibility of decomposing a system in many different ways is a potential source of semantic richness, since each decomposition can embody a different meaning. If one allows relations of higher ordinality than two, i.e., considers not only graphs but hypergraphs, in which relations can be triadic, tetradic, etc., an even greater number of decompositions is possible. For example, four elements have 114 different hypergraph structures (Zwick, "Overview of Reconstructability Analysis"), and thus a tetradic symbol could have as many as 114 different meanings. If relations have directions, there are still more. A symbol consisting of ten elements could in principle have a very large number of structural decompositions and meanings. If one restricts oneself to the much smaller subset of "partitions" in which every element appears in only one substructure, this subset is still quite large. Or, if one restricts oneself to only to graphs, i.e., to structures having only dyadic links, this subset is also large. Table 5.1 just gives a very small hint of this combinatorial explosion, and only samples the sequences that appear in the Chinese literature for the Five Agents.
72. For example, the Tree in Figure 5.1 has only twenty channels, but when channels are correlated with the twenty-two Hebrew letters, two more channels are required; usually these are either Wisdom-Judgment and Understanding-Benevolence or Eternity-Kingdom and Glory-Kingdom.

The Diagram, by comparison, is simpler. The Diagram is built around a composite of the Two Forces and the Five Agents. Chinese philosophy did not utilize a symbolism of seven, although the union of the Two Forces and Five Agents was conceptualized early in Chinese thought, and the seven elements are referred to as a whole by Zhu Xi.[73] Note that this heptad does not parallel the *Sefirot* of Construction, nor does it parallel the seven vertical levels of the Tree.[74] In the language of systems theory, this composite exemplifies Simon's idea that complexity is often achieved by joining together stable subassemblies[75]; this also illustrates von Bertalanffy's[76] notion of "progressive systematization." To this heptad, circles I, IV, and V are added, these additions being already present in the Daoist precursors of the Diagram. Interestingly, it is precisely the addition of these three circles that establishes the near isomorphism of the Diagram with the Tree.

Because of its symbolism of ten and multiple ways of defining substructures and because the *Sefirot* constitute a homogeneous set of elements, the Tree is more integrated than the Diagram. The channels between the *Sefirot*, associated with the Hebrew letters, were often a significant part of the symbolism. In contrast, explicit relations between elements of the Diagram show up only within the Five Agents. There are no links between an individual Force and an individual Agent or between a Force or Agent and circle IV or V, nothing analogous to the direct relations between Wisdom and Benevolence or between Beauty and Foundation. The Diagram looks like a set of unconnected sub-

73. Fung (*History of Chinese Philosophy*, 547) gives the Zhu Xi quote. The linkage of the Two Forces and the Five Agents was an ancient one, not an innovation of Zhu Xi. Berling notes that "Yin and Yang and the Five Agents had first been united in a primitive cosmology by one Tsou Yen, two hundred years before the Han" dynasty of 200 BCE—220 CE (Berling, *Syncretic Religion*, 21). This heptadic grouping notwithstanding, an explicit symbolism of seven was generally absent from Chinese thought. By contrast, seven is ubiquitous in Western symbolism.

74. These seven levels were connected in occult Kabbalah to the seven chakras.

75. Simon, *Sciences of the Artificial*, chapter "The architecture of complexity." Because the Diagram was constructed from these subassemblies it was not readily decomposable in other ways; by comparison, the Tree was not a fusion of preexisting subassemblies, so the variety of its structural representations was greater. Simon argues that most systems are 'nearly decomposable' that is, if one partitions them into disjoint substructures, not a great deal is lost. In these terms, the Diagram is much more 'nearly decomposable' than the Tree. Or, to use another systems term, the Tree is more "holistic" than the Diagram, structurally speaking.

76. von Bertalanffy, "General System Theory."

structures. Nonetheless, Yin and Yang Agents are obviously related to the Yin and Yang of the Two Forces, although the Diagram does not display these relations explicitly. Zhou Dunyi writes, "The true (principle) of that which has no Pole, and the essences of the Two (Forces) and the Five (Agents) unite (react) with one another in marvelous ways, and consolidations ensue."

Meditative Uses

The Diagram traces back to a Daoist symbol used to guide meditation. Needham suggests that "it originated with Chen Tuan (d. +969), the famous Wu Dai expositor of the *Yijing*."[77] The elements of Chen Tuan's diagram are listed in Table 5.4. As a meditation guide,[78] it was read from the bottom up rather than from the top down, and served spiritual practice rather than philosophical theory.

TABLE 5.4. Labels of the Diagram of Chen Tuan[79]

Circle I	Transmuting the Spirit so That It May Revert to Vacuity; Reversion to the Ultimateless
Circle II	Taking from *Kan* to Supplement *Li*
Five Agents (III)	The Five Forces Assembled at the Source
Circle IV	Transmuting the Essence so as to Transform It Into the Vital Force; Transmuting the Vital Force so as to Transform It Into the Spirit
Circle V	Doorway of the Mysterious Female

The Diagram commentaries reflect Daoist influence in the alchemical reference of circle IV, in the Five Forces, and in the reference

---

77. Needham, *Science and Civilization in China*, 467.

78. In their meditative context, circle IV represented the transformation of essence (whose material form is semen) into breath into spirit; component III, the "lesser circulation" of the "Five Breaths;" *Kan* and *Li*, the "grand circulation" of the breath, leading to circle II, spiritual consciousness; ending finally in circle I, the return of spirit to nonbeing (*Hsu* or *Wuji*) (Fischer-Schreiber, *Shambhala Dictionary of Taoism*, Shambhala, 14–16; see also the more extensive discussions of Chang). This progression roughly resembles (but certainly not in detail) the levels of the human soul in Kabbalah (see footnotes 37 & 38). In this connection, an eighteenth century diagram on Daoist meditation given by Richard Wilhelm in *Secret of the Golden Flower* is similar to the Diagram and its precursors, and in fact looks even more like the Tree.

79. Fung, *History of Chinese Philosophy*, 441.

to the "Ultimateless" of circle I. Zhou Dunyi reinterpreted this symbol cosmologically and morally. Although meditation was practiced by Neo-Confucians[80] as part of self-cultivation, the Diagram does not seem to have been linked to this practice. The *Sefirot* were also used for meditation,[81] and a bottom-up reading of the Tree sometimes characterized such uses.[82] So both Chinese and Jewish symbols were read upwards to guide meditative practice and downwards to represent cosmological or divine unfolding. Both symbols offered a hierarchical scheme for the soul (spirit, mind). Both characterized the bottom element as female, but not in the abstract and straightforward sense of Yin and Understanding. The femaleness of circle V is "mysterious" and a "doorway," just as "the last *Sefirah* is for man the door or gate through which he can begin the ascent up the ladder of perception to the Divine Mystery."[83]

As for meditative practice itself, the two traditions were quite different. Generally the personal experiences of the Kabbalists were not made public, but their meditation practices that we know of were centered in the names and attributes of God and focused on words and letters which were conceptualized, visualized, or vocalized. In contrast, Daoist meditation employed the circulation of vital energies strongly coupled to breath, sensation, and awareness. The Kabbalist Abulafia, however, did also make use of breathing exercises.[84] A discussion of Daoist and Kabbalist spiritual practices that asserts a deep similarity of the Diagram and the Tree is given by Yudelove.[85]

---

80. Meditation, as "self-cultivation" was practiced by both Zhou Dunyi and Zhu Xi (Julia Ching, in Wing-tsit Chan *Chu Hsi and Neo-Confucianism*, 282).

81. Kaplan (*Sefer Yeszirah: The Book of Creation*, xi) asserts that the *Sefer Yetsirah* is a meditation manual, but such a characterization is clearer for the *Shaarey Orah* of Joseph Gikatila (1248–1323), translated into Latin by Paul Ricci in 1516 and printed in Hebrew forty-five years later (Kaplan, *Meditation and Kabbalah*, 127).

82. Kaplan (*Meditation and Kabbalah*, 118, 121, 125, 132) asserts this, referring to the Kabbalist books of *The Gate of Kavanah of the Early Kabbalists* (*Shaar HaKavanah LeMekubalim HaRishonim*; late 1100's), probably authored by Rabbi Azriel of Gerona, and *Shaarey Orah* of Rabbi Joseph Gikatila. See also Scholem, *On the Kabbalah and Its Symbolism*, 126. Abulafia also hinted at the ascent through the "ladder of the *Sefirot*" (Kaplan, *Meditation and Kabbalah*, 78–79).

83. Scholem, *Kabbalah*, 112.

84. Kaplan, *Meditation and Kabbalah*, 79.

85. Yudelove, *The Tao & The Tree of Life*.

ON THE POSSIBILITY OF INFLUENCE

Since the "null hypothesis" in comparing a Chinese and a Jewish symbol must be difference, it is similarity that requires explanation. It would be simplest to assume that the symbols developed independently and commonalities reflect religious or philosophical universals of thought and experience. But the possibility of intercultural contact should also be examined, especially since diagrams travel light. To consider the possibility of influence, some relevant dates are worth reviewing. The essay of Zhou Dunyi and the commentary of Zhu Xi were written in the eleventh and twelfth centuries, respectively. The similar symbol of Chen Tuan is said to date from the tenth century, and Needham writes that a similar structure occurs even earlier in an eighth century Daoist book.[86] While Chen Tuan's symbol[87] was the same as Zhou Dunyi's Diagram, the eighth century structure[88] was different from it.

The doctrine of *Sefirot* goes back at least to the pre-Kabbalistic *Sefer Yetsirah* (third to sixth century), and the decad as central to creation derives from still older Jewish and Gnostic sources.[89] The *Sefer Yetsirah* referred to ten *Sefirot*, but a full metaphysical theory of the *Sefirot* was not yet explicitly developed. In the *Sefer Bahir* of Provence (and other texts of the thirteenth century), Foundation was assigned to the seventh place. It was moved to the ninth position in writings of the later Kabbalist school in Gerona, Spain.[90] As for the Tree itself, Scholem indicates that it dates at least to the fourteenth century. At the latest, it appears as the frontispiece of the Latin translation by Paul Ricci published in 1516 of

---

86. Needham (*Science and Civilization in China*, 467) gives the title as: *Shang Fang Ta Tung-Chen Yuan Miao Ching Thu* (*Diagrams of the Mysterious Cosmogenic Classic of the Tung-Chen Scriptures*).

87. Fung (*History of Chinese Philosophy*, 441) gives only the commentary but not the structure. Chang, (*Creativity and Taoism*, 164ff.) gives both; these are reproduced in *The Shambhala Dictionary of Taoism* (p. 15). The small circle on the bottom of the Five Agents is omitted there.

88. Fung (*History of Chinese Philosophy*, 439) also provides the structure and gives its title as *Diagram of the Truly First and Mysterious Classic of the Transcendent Great Cave*.

89. Idel, *Kabbalah: New Perspectives*, 112–22.

90. Scholem indicates that Foundation in the *Bahir* preceded Eternity and Glory (*Kabbalah*, 107). Yet a different order is given by Aryeh Kaplan in his translation and commentary (*Bahir*, 117): Glory (6), Foundation (7), Beauty (8), Eternity (9), Kingdom (10).

the *Shaarey Orah* of Joseph Gikatila (1248–1323), a translation which contributed to the development of Christian and occult Kabbalah.

Thus the doctrine of the *Sefirot* and the symbolism of ten appear to be earlier than the Diagram and its Daoist precursors, but the canonical structure of the Tree appears to be later. Since it is not known when *Sefirotic* diagrams first came into being, there is no solid chronological basis on which to build hypotheses of contact or influence from one culture to another. If one tried to construct such a hypothesis, the known dates of appearance of the symbols would argue for a Chinese to Jewish direction, and this might be supported by the fact that a permanent Jewish settlement was established in Kaifeng in the eleventh century, which was then the capital city for the Song dynasty and China's principal cultural and commercial center.[91] Jews are thought to have arrived between 960 and 1126 perhaps from Persia (or Yemen, Bokhara, or even India); the first synagogue was built in 1163. There were earlier visits of Jews to China. A possible—later—link on the European side might have been the Jewish community of the Italian city of Ancona, which in the thirteenth century had trade relations throughout the Mediterranean and "to major hubs for Asian Commerce like Cairo and Baghdad, Constantinople and the Black Sea ports."[92]

On the other hand, the appearance of the structures themselves might suggest a Jewish to Chinese direction. The Tree is highly integrated compared to the composite Diagram. One is struck in the Diagram with the *ad hoc* quality of circles I, IV, and V, which are added to the canonical Two Forces and Five Agents. A symbol whose structure is partially *ad hoc* is more likely to have been influenced by one whose structure is well integrated rather than the reverse. Nonetheless, it is hard to imagine the availability of a version of the Tree to tenth century (or earlier) Daoists, since the Tree seems to have been articulated only much later.[93]

---

91. Michael Pollak, *Mandarins, Jews, and Missionaries*, chapter 13: Beginning of Judaism in China. Pollak sees evidence that the Kaifeng community maintained contact with extra-Chinese Jewish centers for at least several generations in the fact that this community was familiar with Maimonidean doctrine.

92. Spence, "Leaky Boat to China," 20–21.

93. A much earlier origin for the Tree has been proposed by Simo Parpola ("Assyrian Tree of Life," 161–208), who argues that the Tree derives from ancient Assyrian "tree of life" symbolism. This radical proposal is best left to scholars of Kabbalah to evaluate, but it seems inconsistent with the very late public emergence of the canonical structure of the Tree. The structural similarities of Assyrian and Kabbalist diagrams are much weaker than the similarity noted here between the Tree and Diagram.

But as there is no historical evidence for influence in either direction, one might turn to the alternative hypothesis of independent convergent development, since the symbolisms of number and form and the macrocosm-microcosm analogy are ubiquitous in traditional religions and philosophies,[94] and represent a universal mode of metaphysical understanding. The Neo-Confucian and Kabbalist traditions both encompass this type of metaphysics. However, this hypothesis does not seem satisfactory either, since it is hard to believe that these commonalities adequately account for the extent of resemblance between the symbols.

SUMMARY

To recapitulate: Structurally, the two symbols reflect an early (non-scientific and pre-mathematical) form of systems thinking. The symbols are nearly isomorphic, i.e., the elements of one map onto those of the other

---

94. To complicate matters further, there is another similar metaphysical symbol, the Hindu Tantrik Sanhkhya Tattva diagram (Rawson, *Art of Tantra*, 182), which has some similarities to the Diagram and the Tree. This symbol depicts "creation" and the downwards transition from unity to multiplicity—and simultaneously—the structure of the "subtle body" and its upwards reintegration by Sadhana. The diagram features male and female columns, beginning with Shiva and Shakti which might be correlated with Yang and Yin and with Wisdom and Understanding. This primary dyad emerges out of or separates within "Brahman without Qualities" and "All-embracing Parasamvit" recalling perhaps *Wuji or Ein Sof*. It descends on the side of Shakti to a cluster of five Kanchukas, possibly paralleling the Five Agents, which are attributes of consciousness or thought and the domain of Maya, illusion. Beneath this, the columns diverge distinctly into male and female Purusa and Prakrti which parallel in erotic imagery (Rawson, *Art of Tantra*, 130) the male and female aspects of circle IV and Foundation. The lowest level of the diagram in the male column consists of the multiplicity of Purusas—"I's" which "believe themselves separate," paralleling the Chinese "myriad things" of circle V and the multiplicity of Kingdom. The Tantrik diagram differs significantly from both Chinese and Jewish symbols in the absence of neutral elements, and there are numerous other differences, but this symbol is clearly of the same "genre" as the Diagram and the Tree.

Scholem (*On the Mystical Shape of the Godhead*, 194–96) in fact compared the representations of the *Sefirotic* world with the yantras (meditation diagrams) of Indian Tantrik religion. He pointed to the similarity between the *Sefirotic* pair of Understanding-Wisdom and Shakti and her male counterpart, but also insisted that the differences between the Tantrik and Kabbalistic symbols were "no less profound than their affinities." Scholem must have been surprised to encounter other similarities as he wrote, "The student [of Heinrich Zimmer's work on these diagrams] will be amazed to discover the Kabbalist symbols of the point and the triangle in these remarkable discussions of Indian material." Borrowing, generalizing, and reversing Scholem's phrase, one can argue that the affinities of the Tree and the Diagram are no less profound than their differences.

and many corresponding elements and relations are similar in meaning or structure. Beyond their graph-theoretic connectivities, both symbols have the same spatial distribution of horizontal polar dyads and vertical hierarchical levels. In both, neutral elements harmonize these polarities or are their source or terminus. If, in the Diagram, Yang and Yin (circle II) are counted as two elements and circle IV as one, there is in fact a 1:1 mapping between the ten elements of the two symbols (but no 1:1 mapping between their linkages). The hierarchy of each diagram closes upon itself, with the first and last elements, primal unity and unfolded multiplicity, closely linked. Both symbols declare the isomorphism of macrocosm and microcosm: they are read downwards as cosmological or theosophical diagrams, but upwards as instruments of spiritual practice. In both symbols, two ideas—positive and negative, the manifest and the unmanifest—are associated with the first element, with the dualism resolved in different ways. The meanings of the first three and last two elements are similar, with sexual generativity implied in elements two and three and element nine. The central portions of both diagrams exhibit two dyads and a neutral harmonizing element. They present benevolence (love, mercy, humanity) and righteousness (justice, rigor) as the primary virtues, and as male and female, respectively. Moral action is referred in both to element nine. Element ten is feminine and represents the consequences flowing from sexual generativity (or moral discrimination) of element nine, namely the material (or behavioral) multiplicity of the world.

Given the many differences between Chinese and Judaic thought in general, and between Neo-Confucianism and Kabbalah in particular, this list of similarities is striking. The purpose of this paper is to call attention to these similarities, which remain to be explained, while noting also the differences between the symbols. The similarities that exist may arise from the presence in Chinese and Jewish thought of universal ideas and modes of thought also prominent in other philosophical and religious traditions; or, there may have been some actual intercultural influence. No attempt has been made here to resolve this question, which will hopefully be the subject of future investigation.[95]

---

95. Acknowledgements: The author is indebted to Anthony Blake for stimulating discussions on religious symbolism, to Joseph Adler and Anne Birdwhistell for their valuable comments on Neo-Confucianism and the Diagram, to Joseph Dan for his observations on the peripheral status of the Tree in Kabbalist thought, and to Irene Eber

APPENDIX: COMMENTARIES ON THE DIAGRAM OF THE SUPREME POLE (TRANSLATED BY NEEDHAM)

The exposition of Zhou Dunyi

(1) That which has no Pole! And yet (itself) the Supreme Pole!

(2) The Supreme Pole moves and produces the Yang. When the movement has reached its limit, rest (ensues). Resting, the Supreme Pole produces the Yin. When the rest has reached its limit, there is a return to motion. Motion and rest alternate, each being the root of the other. The Yin and Yang take up their appointed functions and so the Two Forces are established.

(3) The Yang is transformed (by) reacting with the Yin and so Water, Fire, Wood, Metal, and Earth are produced. Then the Five *Qi* diffuse harmoniously, and the Four Seasons proceed on their course.

(4) The Five Agents (if combined, would form), Yin and Yang. Yin and Yang (if combined, would form) the Supreme Pole. The Supreme Pole is essentially (identical with) that which has no Pole. As soon as the Five Agents are formed, they have each their specific nature.

(5) The true (principle) of that which has no Pole, and the essences of the Two (Forces) and the Five (Agents) unite (react) with one another in marvelous ways, and consolidations ensue. The Dao of the heavens perfects maleness and the Dao of the earth perfects femaleness. The Two *Qi* (of maleness and femaleness), reacting with and influencing each other change and bring the myriad things into being. Generation follows generation, and there is no end to their changes and transformations.

(6) It is man alone, however, who receives the finest (substance) and is the most spiritual of beings. After his (bodily) form has been produced, his spirit develops consciousness; (when) his five agents are stimulated and move, (there develops the) distinction between good and evil, and the myriad phenomena of conduct appear.

---

for helpful assistance with Chinese philosophical ideas and terminology. Anonymous reviewers of past drafts of this paper have made useful comments, and the author is also grateful for the valuable suggestions of David Rounds, the editor of *Religion East and West*, in which a shorter version of this article has been published. The assertions of this paper are of course the responsibility only of the author.

(7) The sages ordered their lives by the Mean, by the Correct, by Love and Righteousness. They adopted ataraxy as their dominant attitude, and set up the highest possible standards for mankind. Thus it was that the 'virtue of the sages was in harmony with that of heaven and earth, their brightness was one with the Four Seasons, and their control over fortune and misfortune was one with that of the gods and spirits.'

(8) The good fortune of the noble man lies in cultivating these virtues; the bad fortune of the ignoble man lies in proceeding contrary to them.

(9) Therefore it is said, "In representing the Dao of Heaven one uses the terms Yin and Yang, and in representing the Dao of Earth one uses the terms Soft and Hard; while in representing the Dao of Man, one uses the terms Love and Righteousness." And it is also said, "If one traces things back to their beginnings, and follows them to their ends, one will understand all that can be said about life and death."

(10) Great is the (Book of) Changes [*Yijing*]! (Of all descriptions) it is the most perfect.

The Commentary of Zhu Xi

(a) The uppermost figure represents that of which it is said, "That which has no Pole! And yet (itself) the Supreme Pole!" It is the original substance of that motion which generates the Yang (force), and of that rest which generates the Yin (force). It should be regarded neither as separate from, nor as identical with, the Two Forces.

(b) The concentric circles in the second figure symbolize motion giving rise to Yang and rest giving rise to Yin. The complete circle in the center symbolizes the substance which does this (equivalent to the circle of the first figure). The semicircles on the left indicate the motion which produces Yang; this is the operation of the Supreme Pole when moving. The semicircles on the right indicate the rest which produces Yin; this is the substance when at rest. Those on the right are the root from which those on the left are produced and vice versa (i.e., Yang generating Yin, and Yin generating Yang).

(c) The third figure symbolizes the transformations of the Yang and Yin forces in union with each other, and thus the generation of the Five

Agents. The diagonal line from left to right symbolizes the transformation of the Yang, and that from right to left symbolizes the unions of the Yin.

Water is predominantly Yin and its place is therefore on the right. Fire is predominantly Yang and its place is therefore on the left. Wood and Metal are modifications of the Yang and Yin respectively, and therefore they are placed to the left and right under Fire and Water. Earth is of mixed nature, therefore it is placed centrally. The crossing of the lines above the positions of Fire and Water indicates that the Yin generates Yang and vice versa. (The order of their generation is indicated by the intersection lines connecting the Five Agents), Water, being followed by Wood, Wood by Fire, Fire by Earth, Earth by Metal, and Metal again by Water, in an endless unceasing round, so that the five *Qi* spread abroad and the four seasons revolve.

(d) The Five Agents all come from the Yin and Yang (Forces). The five different things (fit into) the two realities without the slightest excess or deficiency. And the Yin and the Yang (go back to) the Supreme Pole (perfectly), neither one of them being more or less elaborate than the other, nor more or less fundamental than the other.

The Supreme Pole is essentially the same as that which has no Pole. Noiseless, odorless, it exists everywhere in the universe. As soon as the Five Agents are generated, they have each their specific natures. Since these *Qi* are different, the tangible matters (which manifest them) are also different. Each sort has its completeness, and this there is no gainsaying.

The small circle below, connected by the four lines with the Five Agents above, indicates that which has no Pole, in which all are mysteriously unified, as indeed again cannot be denied.

(e) The fourth figure represents (the operations of the *Qi* of Yin and Yang exhibited in) the principles of (heavenly) maleness and of (earthly) femaleness which pervade the universe, each having their own natures, but (both going back to) the one Supreme Pole, (as indicated by the reproduction of the original circle).

(f) The fifth figure represents the birth and transformation of the myriad things in their sensible forms, each of which has its own nature. But, (as indicated again by the reproduction of the original circle), all the myriad things go back to the one Supreme Pole.

6

Religion and the System of Meaning

DAVID J. KRIEGER

ABSTRACT

Science has always asked questions about order in nature, human existence, and society. Nature manifests many different forms of order, for example, galaxies, planetary systems, geological formations, meteorological phenomena, chemical compounds, organisms, tissues, cells, molecules and atoms. Human society and culture also manifest order, for example, social systems, political institutions, languages, works of art, and scientific theories. The paradigm of self-organizing systems attempts to approach the problem of emerging order by developing a theory that views reality, both natural and cultural, as a systemic phenomenon. There may be said to be three levels of emergent order: mechanical systems, biological systems, and systems of meaning. A system of meaning is a human society or culture. It operates in order to construct meaning. The boundary of the system is at the same time always a border between meaning and meaninglessness. The uttermost boundary of the system of meaning builds a common horizon of meaning, a worldview, shared values, and a basic understanding of what is real, true, good, and beautiful. It is this "life world horizon" that makes mutual understanding, cooperative action, and a shared life within society possible. Everything outside this uttermost boundary of shared meaning and value simply does not "fit" in the world. It is marked as "impossible," "evil," "abnormal," "irrational," "barbaric," and so on. The way in which a system of meaning differentiates itself from meaninglessness, chaos, and disorder constitutes the function of religion. Ideological conflict is a result of functionally equivalent communication on the religious level. Religion is important because boundaries between self and other, meaning and meaninglessness must be drawn in every instance of systemic and subsystemic order. Drawing these boundaries requires a specific form of communication, a "boundary" discourse that may be characterized as religious communication. The pragmatics of boundary discourse are proclamation, narrative repetition, ritual representation, temporal orientation towards founding events, and inclusion/exclusion

WHAT MAY BE REFERRED to as the "paradigm of self-organizing systems" is not a single theory, but a framework of theories and basic concepts from a variety of disciplines. The components of this framework consists of systems theory, information theory, cybernetics, the theory of autopoietic systems in biology, synergetics, the theory of dissipative structures, theories of biochemical evolution, deterministic chaos, and related disciplines.[1] All these different theories make use of common concepts, for example "system," "complexity," "control," "emergence," "discontinuity," "code," and "information." These theories also have a common goal, to discover general rules governing the emergence and maintenance of order. This is, of course, not a new goal for human science. Ever since its beginning, science has asked questions about order and regularity in nature, human existence, and society. Nature manifests order, for example, in galaxies, planetary systems, geological formations, meteorological phenomena, chemical compounds, organisms, tissues,

---

1. On the "paradigm of self-organizing systems" see Krohn/Küppers/Novotny, *Emergenz: Die Entstehung von Ordnung, Organisation und Bedeutung*; and Dalenoort, *Paradigm of Self-Organization*. General systems theory was first introduced by von Bertalanffy, *General System Theory*. Information theory and cybernetics have been described by Ashby, *Introduction to Cybernetics*; Wiener, *Cybernetics: or Control and Communication in the Animal and the Machine*; and von Foerster, *Observing Systems*. The theory of autopoietic systems in biology has been proposed by Maturana and Varela, *Tree of Knowledge: The Biological Roots of Human Understanding*. For synergetics see Haken, *Synergetics: An Introduction*; *Information and Self-Organization*; and for the theory of dissipative structures see Prigogine, *Order out of Chaos: Man's New Dialogue with Nature*. Important contributions in the area of biochemical evolution have been made by Eigen, *Organization of Matter*. The description of deterministic chaos has been undertaken by Lorenz, *Deterministic Nonperiodic Flow*. Other literature can be found in the bibliography.

cells, molecules and atoms. Order is not unique to the realm of nature. Human society and culture also manifest order. There are social systems, political institutions, languages, works of art, and scientific theories, all of which exhibit order. The paradigm of self-organizing systems attempts to view all these phenomena under the aspect of systemic organization.

A system can be defined as a functional set of relations among selected elements. From the point of view of systems theory, whatever kind of order we may be speaking of, whether in nature or society, we are talking about elements integrated into a whole according to certain rules. A galaxy is made up of stars, planets, gases, molecules, water is made up of hydrogen and oxygen, an organism is made up of cells. All of these are combinations of elements in certain relations, that is to say, they are systems. The same may be said of social and cultural phenomena. Social institutions and cultural phenomena are made up of actors, artifacts and signs that are related according to certain rules. The theory of systemic order opens up a unifying perspective from which we may view reality, whether natural or social, as a systemic phenomenon.

One of the striking effects of a general theory of self-organizing systems lies in its power to unite perspectives drawn from both the natural and the social sciences. Of course, there have been previous attempts to bring the sciences of nature and of culture under a single methodological perspective. Positivism, for example, has long placed the scientific status and legitimacy of the humanities in question. The so-called "soft" sciences have reacted to this criticism in two different ways. One well known reaction has been to embrace positivism and attempt to reconstruct the humanities on concepts and methods drawn from the natural sciences. Only hypotheses that can be empirically falsified count as scientific and thus as meaningful knowledge. The reaction to positivism has been the attempt to develop a form of knowing specific to the study of culture. The methodology that filled this demand has come to be known as "hermeneutics." Hermeneutics is the science of interpretation and understanding, originally of sacred, legal, and literary texts. According to hermeneutics, society and culture can be investigated by means of specific techniques of interpretation and critique and not by means of empirical falsification. What the social sciences discover is therefore not a mathematically expressible description of the behavior of physical bodies, but the ways in which the human mind reveals and conceals knowledge about the world and about itself. To unveil knowledge that has been hidden is not only a critique of the deceptions and illusions that are disguising truth, but it is also a liberating and interpretive act grounded in subjective insight and decision. Critique is therefore always to a certain extent arbitrary and for this reason itself subject to criticism. Critique as a form of scientific inquiry is always self-critique. Self-critique is at once enlightening and liberating. Hermeneutics was the answer to positivism both inside and outside the social sciences. A way of knowing that enlightens and liberates incorporates the very meaning of rationality. Enlightened self-criticism with its accompanying subjectivity and self-reference became scientific methodology with its own claim to universal validity. Thus was born the idea of a critical science capable of disclosing the hidden ideological commitments of positivism. In recent decades, this tradition in the social sciences has been characterized by a radical critique of universalism. General theories as such have become suspect. Under the name of "postmodernism,"[2]

2. The controversy about "erklären" and "verstehen" (explanation and understanding) was initiated in German sociology by Max Weber and has been carried on in various forms up to the present day in the work of Jürgen Habermas. For a good discussion of the issues involved see Apel *Understanding and Explanation*.

For postmodern theory see the works of Jean-François Lyotard (especially *The Postmodern Condition*) and Jacques Derrida (for example: *Margins of Philosophy*). Postmodernism as a theoretical movement arose from post-structuralism and deconstruction in the 1960s. Mary Klages offers the following concise definition of postmodernism: "The ways that modern societies go about creating categories labeled as 'order' or 'disorder' have to do with the effort to achieve stability. Francois Lyotard . . . equates that stability with the idea of 'totality,' or a totalized system (think here of Derrida's idea of 'totality' as the wholeness or completeness of a system). Totality, and stability, and order, Lyotard argues, are maintained in modern societies through the means of 'grand narratives' or 'master narratives,' which are stories a culture tells itself about its practices and beliefs. A 'grand narrative' in American culture might be the story that democracy is the most enlightened (rational) form of government, and that democracy can and will lead to universal human happiness. Every belief system or ideology has its grand narratives, according to Lyotard; for Marxism, for instance, the 'grand narrative' is the idea that capitalism will collapse in on itself and a utopian socialist world will evolve. You might think of grand narratives as a kind of meta-theory, or meta-ideology, that is, an ideology that explains an ideology (as with Marxism); a story that is told to explain the belief systems that exist. Lyotard argues that all aspects of modern societies, including science as the primary form of knowledge, depend on these grand narratives. Postmodernism then is the critique of grand narratives, the awareness that such narratives serve to mask the contradictions and instabilities that are inherent in any social organization or practice. In other words, every attempt to create 'order' always demands the creation of an equal amount of 'disorder,' but a 'grand narrative' masks the constructedness of these categories by explaining that 'disorder' *really is* chaotic and bad, and that 'order' *really is* rational and good. Postmodernism, in rejecting grand narratives, favors

critique of universalism and a special hermeneutics of the particular have become very influential.

Even if the methodological battle between *erklären* and *verstehen* seems to have exhausted itself or simply been left behind while researchers turn to other interests, there is an uneasy truce between the natural and social sciences as long as the social sciences satisfy themselves with postmodern and deconstructive approaches. What does the radical program of deconstruction lead to if not to the Humpty Dumpty question: How do we put the pieces back together again? Postmodern critique of general theories is doubtlessly as important a contribution to human self-understanding as is the detailed studies of micro-cultural phenomena that postmodern social science has brought forth. In both areas, the program of deconstruction has made important contributions. Nevertheless, it may be time to ask if the rejection of general theories by the social sciences does not block further theoretical development.

A general theory of systemic order provides an opportunity to step back from the methodological deadlock between the natural and the social sciences and view phenomena of order from a new perspective. The mind may be seen as a system of meaning obeying laws of its own. These laws require that meaning arise within an encompassing and unifying horizon of possibilities and impossibilities that can only be articulated in narrative form. Narratives describing the world and what does not belong to the world have been termed myths. Myths are stories about the world as a whole, about what counts as real, true, valuable, and beautiful and what does not. Myths are not restricted to traditional religion, but play a role in secular and scientific worldviews as well. The nature of narrative is such that there can be no absolutely true story. No story can tell everything. Only certain possibilities are allowed. Something is always excluded from the system. The fact that narratives cannot tell everything, but only what fits at a certain point into the storyline, can have positive as well as negative consequences. Postmodern critique is well aware of the deceptive strategies of the mythic construction of meaning. Under the influence of this insight, contemporary social science tends to regard general theories as myths in the negative sense of untrue stories, that is,

---

'mini-narratives,' stories that explain small practices, local events, rather than large-scale universal or global concepts. Postmodern 'mini-narratives' are always situational, provisional, contingent, and temporary, making no claim to universality, truth, reason, or stability" (http://www.colorado.edu/English/ENGL2012Klages/pomo.html).

ideologies. It has become recognized in recent decades, however, that story telling plays a very important part in the communicative construction of meaning. If a reappraisal of the possibility of a unified science of order is to occur, then the social sciences must overcome their fear of positivism and their rejection of universalism and take a new look at the theoretical resources the natural sciences now have to offer.

The general theory of self-organizing systems attempts to describe the basic characteristics of systemic order. It tries to show what all systems have in common. In the following we will briefly look at some the major characteristics of all systems as such, then turn specifically to a description of meaning systems, and finally, address the question of religion and the function of religion in society.

Every System Has a Principle of Organization, a "Code," that Fulfils Three Functions: Selection, Relationing, and Steering

One of the basic assumptions of general systems theory is that almost everything may be viewed as a system. A system (from Greek *to systema*) is by definition a composition of elements. A table, for example, consists of a top and of legs. In order to have a table one has first to *select* the top and the legs out of all possible things in the world. Secondly, one has to put these elements into certain *relations* with each other. The top has to sit on the legs and not the other way round. This may be termed "relationing". Finally, the relationing of the elements can be considered to control or *steer* the function or operation of the system. Every system has a specific purpose or function that it fulfills. The table has the function of, let us say, providing a working space at middle body height. Systems are always functional entities insofar as they always perform some kind of operation. This means that systems do not merely exist, they act. There is a system, therefore, only when elements are *selected*, *related* and *controlled or steered* in such a way that a function is fulfilled. The principle of organization that selects, relates, and steers a system may be called a "code." From the point of view of general systems theory, every system is organized by a code that fulfills the three functions of selection, relationing, steering.

Every System Is Based upon a Difference between Itself and Its Environment; This Difference Is Constitutive for the System

When a system comes into being, then there is much in the world that is excluded from it. The very notion of selection implies not only inclusion, but also exclusion. To select is to choose certain possibilities out of a variety of options and thus not to choose others. This means that a system is always made up of less than all possible elements and combinations thereof. Selection, therefore, implies exclusion. Everything that is excluded from a system is called the "environment." Since a system is constituted by selection, every system will have its environment. In other words, a system can be a system only *because* it is distinguished from the environment. If a system is not different from the environment, there is no system. If the difference between the system and the environment cannot be maintained, the system disintegrates and disappears. A system, therefore, is based upon a difference.

It is important to note with regard to what follows that the concept of "difference," like the concept of "code," is a central concept in semiotics. When Saussure describes "language" as a system of differences and when Bateson defines "information" as a difference that makes a difference, they are pointing to an aspect of meaning that can very well be understood from the perspective of systems theory. The concept of "difference" plays a similar foundational role in systems theory to the role it plays in semiotics and in much postmodern thought as well.

Every System Constructs Its Own Elements

Just as a system is always made up of less than what is possible, it is also always more than a mere a collection of parts. A system is not a disordered pile of things that previously were lying around in the environment. A table, for example, consists of a top and of legs. Before there were tables, there were no such things as "tops" and "legs" lying about in the world. There were pieces of wood, slabs of stone, etc, but there were no "tops" and no "legs," that is, not until the code that organizes the table selected and related certain wooden, stone, or whatever things for the purpose of providing, let us say, a working space at middle body height. The code, therefore, can be said to have *constructed* the elements of the system out of things that were lying around in the environment. This principle has very important and far-reaching consequences. It means, for example, that the elements of a system are not things or entities in the common or even in the philosophical sense, but instead, they are functions. A system does not consist of things at all, but of elements that are related in such a way that they fulfill a certain function. A top and legs are not things that exist somewhere in the world, they exist only in tables. If there were not tables, there would be no tops and legs. Furthermore, it is important to note that just about anything and everything can function as a tabletop or as the legs of a table. If a person, for example, goes down on hands and knees and holds this uncomfortable position long enough, he or she can serve as a table. Even water, as Eskimos might know, can serve the function of a table. A tabletop and legs are therefore not substances—it doesn't matter what sort of things they are made of—but functional entities. What matters for a system is how something functions. A tabletop not a substance, it is whatever functions as a tabletop. Although the environment puts constraints on functionality, elements are constructed by the code of a system and not by the environment. If general systems theory claims that everything can be viewed as a system, then it is proposing a view of reality as a functional totality and not as the sum of all things.

Functionality can also clearly be seen in the case of organic systems. A living being, let us say, consists of cells, organs, tissue, skin, bones, stomach, liver, heart, etc. Cells and organs obviously do not lie about in the environment until someone like Dr. Frankenstein comes along and puts them together in order to make a living being. The organic system, that is, the genetic code that organizes the biological system constructs its own elements. The way in which the genetic code programs the development of cells and organisms is extremely complex and not at all comparable to the way in which a drawing might be said to "program" the building of a table. Nonetheless, from the point of view of a general theory of systemic order, it may be said that both physical and biological systems consist of elements that have been selected and related into a functioning totality. In both cases the processes of selection, relationing, and steering may be said to construct the elements of the system.

The principle that a system constructs its own elements also clearly applies to semiotic systems, for example, a language. The human voice can produce an almost infinite number of sounds, but only very few are selected as words in a specific language. The typical "th" sound of the English language plays no significant role in German. The code of the

German language has selected only certain possible sounds that the human voice can utter to become elements of a spoken language. Just as a table does not need to be made of wood or metal, but can consist of almost anything, a language does not need acoustic material. Gestures, written signs, and objects of all sorts can serve the same function as verbal signs. A word in a language, therefore, is not a thing or a substance, but—as Saussure noted—a function within a differential system of signs. It is the system, that is, the semiotic code that constructs the significant elements of which a language consists. The principle that every system constructs its own elements provides the theoretical foundation for the school of thought that has come to known as "constructivism."

Constructivism is one of the most interesting developments in recent years.[3] Although it is perhaps best known as a theory of knowledge, there are constructivist ontologies and constructivist theories in psychology, pedagogy, and other domains of the social sciences. Constructivism traces its roots back to Greek skepticism. It shares the skeptical conviction that knowledge of an external reality beyond the mind is not possible. Despite many convincing arguments for the assertion that knowledge is a construction of the mind and not a mirror of reality, constructivism is unable to explain what the mind is, why the mind constructs reality the way it does and how it does this. Cognitive constructivism reduces meaning to recursive operations of nerve impulses in the brain. Social constructivism views meaning as a product of the social interaction of individuals. If meaning, however, is constructed by social interaction, which in turn is constructed by intentional individual actions, how can these individual actions be meaningful? If meaning is a result of action,

3. On constructivism see the works of Heinz von Foerster, Ernst von Glasersfeld, and Humberto Maturana. For the current discussion see the various issues of *Foundations of Science The official Journal of the Association for Foundations of Science, Language and Cognition*, Kluwer; as well as the wealth of material at the Radical Constructivism website: http://www.univie.ac.at/constructivism/; and at the Principia Cybernetica website: http://pespmc1.vub.ac.be/DEFAULT.html

A. Riegler defines constructivism as follows: "Constructivism is the idea that we construct our own world rather than it being determined by an outside reality. Its most consistent form, Radical Constructivism (RC), claims that we cannot transcend our experiences. Thus it doesn't make sense to say that our constructions gradually approach the structure of an external reality. The mind is necessarily an epistemological solipsist, in contrast to being an ontological solipsist who maintains that this is all there is, namely a single mind within which the only world exists. RC recognizes the impossibility of the claim that the world does not exist." (Riegler, *Towards a Radical Constructivist Understanding of Science*, 1.)

how can it at the same time be the cause thereof. Cognitive constructivism cannot explain why there is anything at all beyond the brain, social constructivism on the other hand cannot explain how there can be intentional acting individuals outside of society. At this point self-organization theory can offer a helping hand. If systems construct their own elements, it is the system of meaning alone that constructs meaning and neither the brain nor human individuals. Instead of cognitive or social constructivism, we may propose the alternative of "semiotic constructivism." Semiotic constructivism is an alternative to cognitive and social constructivism, because it considers meaning a level of emergent order and not an effect of non-meaningful causes, whether they be nerve impulses in the brain or the pre-social and thus pre-rational individual actors. On the basis of the general principle that systems construct their own elements, a theory of human culture and society can be developed that unites systems theory, constructivist theories, and semiotics into a "semiotic constructivism".

Every System Is in One Way or Another Self-Referential; That Is, It Refers Its Operations to Itself

If it can be said that every system constructs its own elements, then it can also be said that every system has a tendency to maintain its own organization and structure, that is, to resist change. The table, for example, tends to resist attempts to use it for some other purpose than that for which it was intended. If someone attempts to use a table as a chair or as a kitchen sink, they won't have much success. The table selects the state of being a table, rather than the state of being a chair, a sink or whatever. A table, it can be said, operates in order to maintain its own organization, that is, its own structure and function. Once the table system has appeared, that is, once a certain form of order comes into being, disorder is less probable than order. There arises an asymmetry between order and disorder. Ashby (1956) refers to this aspect of systemic order as the "principle of self-organization." Self-maintenance implies that a system operates in some way upon itself. Such self-referential operations are a prerequisite for systemic order and may be termed the "minimal" form of self-reference. What does it mean to say that the operations of a system "refer" to itself? The idea of self-reference derives from the controlling or steering function of the code. As we have seen, every system is organized by a code that has the functions of selection, relationing, and steering. The code selects certain elements, relates them in certain ways so that the system operates to fulfill a certain function. In terms of classical Western philosophy the code may be said to be the *eidos* or essence of the system, that is, it is that which makes a thing to be what it is. At least in this very minimal sense of structure maintenance, the operations of any system can be termed self-referential.

For an adequate understanding of self-reference it is the steering function of the code that is important. Selection and relationing are "steered" by the code such that a function is fulfilled. The analysis and description of the steering function in systemic organization is usually the task of cybernetics.[4] Insofar as every system is organized by a code that selects, relates, and steers operations, every system can be described from the point of view of cybernetics. A general theory of systemic order is therefore also a general cybernetics. This extends the notion of cybernetics beyond its normal scope. Nevertheless, the moment that reality as a whole is viewed in terms of systemic order, everything whatever can also be viewed from the point of view of cybernetics. For example,

---

4. Cybernetics was introduced by Wiener in *Cybernetics: or Control and Communication in the Animal and the Machine*. Current discussion can be found at the Principia Cybernetica website: http://pespmc1.vub.ac.be/DEFAULT.html, where "cybernetics" is defined as follows: "Norbert Wiener, a mathematician, engineer and social philosopher, coined the word 'cybernetics' from the Greek word meaning steersman. He defined it as the science of communication and control in the animal and the machine. Ampere, before him, wanted cybernetics to be the science of government. For philosopher Warren McCulloch, cybernetics was an experimental epistemology concerned with the communication within an observer and between the observer and his environment. Stafford Beer, a management consultant, defined cybernetics as the science of effective organization. Anthropologist Gregory Bateson noted that whereas previous sciences dealt with matter and energy, the new science of cybernetics focuses on form and pattern . . . [Cybernetics can also be defined as (DK)] an interdisciplinary approach to organization, irrespective of a system's material realization. Whereas general systems theory is committed to holism on the one side and to an effort to generalize structural, behavioral developmental features of living organisms on the other side, cybernetics is committed to an epistemological perspective that views material wholes as analyzable without loss, in terms of a set of components plus their organization. Organization accounts for how the components of such a system interact with one another, and how this interaction determines and changes its structure. It explains the difference between parts and wholes and is described without reference to their material forms. The disinterest of cybernetics in material implications separates it from all sciences that designate their empirical domain by subject matters such as physics, biology, sociology, engineering and general systems theory. Its epistemological focus on organization, pattern and communication has generated methodologies, a logic, laws, theories and insights that are unique to cybernetics and have wide-ranging implications in other fields of inquiry."

although we don't usually think of a table as a cybernetic system, it does steer its operations in the minimal sense of self-referentially in order to maintain its own structure.

The theory of cybernetic or self-steering systems usually describes systems that are dynamic in a much more obvious way than a table. The dynamics of a cybernetic system can be defined as the way in which outputs of the system become inputs back into the system, thus creating a circular causality. The circular causality of a cybernetic system has the effect of maintaining the system in a certain state or "reference value." Cybernetic systems are self-steering systems, since the operations of the system function as inputs back into the system. The system reacts to its own operations such that it can maintain its reference value. A typical cybernetic system is a thermostat. Indeed, the thermostat is one of the most often cited examples of a cybernetic machine. When the temperature in a room falls below a certain point, the thermostat registers this event in the environment as relevant information and turns on the heating unit. When the air temperature rises because of the output of the heater, the thermostat registers this information and turns the heater off. The output of the system thus becomes an input back into the system in a causal circle of operations. In the case of the heating system, we see that it operates in order to maintain a certain state that is defined by the code as a reference value, for example, normal room temperature. In the case of a table, it can hardly be said to be self-steering in this sense of the word, nonetheless even a table can manifest a minimal self-reference, for it operates to maintain itself as a table.

Organic systems have their own form of self-reference and accordingly their own cybernetics. The kind of self-reference peculiar to living systems is called "autopoiesis". Autopoiesis (from Greek "auto" = self and "poiein" = produce) literally means "self-producing." Organic systems are self-producing systems, that is, they operate not in order to change some state in the environment, for example, the temperature in a room, but in order to continue their own operations.[5] The cybernetics of living

---

5. The theory of "autopoiesis" was proposed by Maturana and Varela (see for example: *The Tree of Knowledge*). A typical definition of autopoiesis is: "the process whereby an organization produces itself. An autopoietic organization is an autonomous and self-maintaining unity which contains component-producing processes. The components, through their interaction, generate recursively the same network of processes which produced them. An autopoietic system is operationally closed and structurally state determined" (from http://pespmc1.vub.ac.be/ASC/AUTOPOIESIS.html).

systems may be termed "2nd order cybernetics." Organic systems have a circular causality in a way that mechanical systems do not. It is not accidental or irrelevant that the output of a thermostatically controlled heating system raises the temperature of a room, but it is accidental and irrelevant for a cow that the heat it and other cows discharge in the stall raises the temperature in the barn. The cows eat, breath, and so on in order to continue their own rumination and breathing, not in order to heat the stall so that the farmer can milk them more comfortably. Organic systems operate exclusively in order to continue their own operations. The system itself is what counts and not any changes that might occur in the environment as a result of what the system does.

Finally, with regard to semiotic systems, the kind of self-reference and the kind of cybernetics are of an altogether different sort. For a system of meaning self-reference takes the form of self-identification by means of communication. Meaning systems operate in order to produce meaning. The production of meaning implies the construction of signs, for it is signs that carry meaning. Signs, however, carry meaning only in and through communication. Meaning systems operate, therefore, as communication systems. The way in which communication steers meaning processes is described by "3rd order cybernetics." 3rd order cybernetics describes the specific operational and informational closure of semiotic systems.

In summary, we may say that all systems have certain common characteristics that constitute them as systems. All systems are organized by a code that selects, relates, and steers the elements of the system such that it differentiates itself from an environment and fulfills a certain function. This implies the construction of system-specific elements, a task that can only be accomplished when the system in some way refers its operations back to itself.

Emergent Order

If all systems have the above mentioned four characteristics in common, what makes systems different from each other. This question is not unimportant, since much confusion has resulted from the tendency to apply concepts, models, and methods, which were developed for the analysis and description of one kind of system, to systems of an altogether different kind. The concept of "information" is a case in point. Information means one thing in the theory of signal transmission (Shannon and

Weaver 1949/1963), something completely different in the theory of living systems (Maturana and Varela 1987) and something different again when used in theories of language and culture (Bateson 1988).

If we wish to distinguish different kinds of systems, it is useful to distinguish between different *levels of emergent order*. "Emergence" may be defined as the non-predictable and non-derivable appearance of structures that are capable of "integrating" previously existing structures. Life appeared non-predictably as an integration of physical and mechanical systems, that is, atomic, molecular, and chemical systems. Life cannot be reduced to mechanical or physical systems or derived from them. Even the simplest biological entity is organized by a code that is sufficiently complex to harness physical processes for the construction of living matter. No physical code can do this. The jump to a higher level of emergent order occurs when a new kind of code, in this case, the genetic code, comes into being. The genetic code has the capacity to use and manipulate systemic order on the physical level in ways that physical systems cannot. In the same way, human cultural systems appeared non-predictably as integrations of living systems, above all central nervous systems, and cannot be reduced to these systems or derived from them. No amount of physics or biology will explain a Mozart symphony or even adequately describe a picnic as a cultural event. Much time and energy has been spent trying to explain meaning, thought, self-consciousness, and cultural phenomena on the basis of the operations of the brain. The brain, however, is a biological system. The operations of the brain are those of a genetically coded organism. The brain is not semiotically coded, even if semiotic coding uses the complexity of the central nervous system in order to emerge as a form of order in its own right. No description of recursive structuring of nerve impulses, no matter how complex, will adequately account for the meaning of a word or explain a language. Meaning cannot be reduced to biology. The elements of a meaning system are signs and neither nerve cells nor electronic impulses. For this reason, it is conceivable that a sufficiently complex machine, a computer for example, could perform the same function as the brain and form the basis for intelligence. Meaning, as almost all religions and many philosophical systems have claimed, is not dependent on life. No system of meaning has stepped back from sacrificing biological life for the sake of meaning, without which life is not worth living. Meaning is a unique level of emergent order that cannot be reduced to organic or anorganic systems.

The emergence of higher levels of order on the basis of lower levels makes it tempting to think of "emergence" as synonymous to "self-organization" and to speak of the evolution of the universe as a self-organizing process.[6] Of course, it can also be claimed that God, or the transcendental ego, or some other omnipotent agency created the universe and everything within it. In all these cases one has simply decided to name the "self" that self-organizes in a certain way. The act of naming is very important for it decides who we are and which of many possible worlds we live in. When a meaning system organizes itself around a certain self-designation, a world of possibilities and impossibilities comes into being. On the level of the system as a whole self-reference is always in some way a "religious" decision. The function of religion in society arises from the necessary self-reference of the system of meaning. General systems theory does not attempt to dictate what religious communication talks about. We may believe in whatever we choose or in whatever chooses us. The theory of self-organizing systems has no prejudice with regard to whom or what we say we are, but attempts to describe the way in which self-reference on the level of meaning arises. For systems that construct meaning, self-reference amounts to making the system itself into something meaningful. The meaning that a system gives itself can be said to be its identity. Throughout history the answer to the question of self-identity has been given (or found) in many different ways, indeed, it can be said that the history of cultures and religions is nothing other than the temporal series of meaningful self-references.

The System of Meaning and 3rd Order Cybernetics

How does a meaning system operate? What is 3rd order cybernetics? In answer to the first question, it can quite simply be said, that the operations of a *meaning system* consist in making something meaningful. A meaning system is not interested in changing the environment or in maintaining metabolic processes, but in producing meaning. If something in the world becomes meaningless, the system will attempt to give it meaning or will expel it into the environment.

---

6. Kaufman, *Origins of Order*.

3rd order cybernetics describes how meaning is produced. In order for something to become meaningful it must be given a name, a designation, a signification, that is, it must be taken up into *semiotic coding*. The elements of a semiotic system are neither machines nor organisms, but signs. This is why meaning systems may also be called *semiotic systems*, for semiotics is the science of signs. Whatever has meaning may be considered a sign of one kind or another. A semiotic system, like all systems, refers its operations to itself and thus is self-referential. Meaningful self-reference is self-designation. To refer to the self, however, implies referring to the non-self, the other, for it draws the boundaries around the system and thus distinguishes the system from the environment. Systemic order, we recall, is based upon difference, and the most important difference for a system of meaning is the distinction between meaning and meaninglessness. This distinction is the primary information for a meaning system, that is, it is information upon which all other information within the system is based. A semiotic system can refer its operations to itself and bring about operational closure only by distinguishing between what belongs to it and what does not belong to it. This distinction is drawn the moment the system designates itself. When a system designates itself, it constructs its "identity." Identity is nothing other than meaningful self-reference. Meaning systems, therefore, are often considered to be "self-conscious," "psychic," or "mental," or however else "we" designate and thus give meaning to "ourselves." The self-reference of a system of meaning is nothing other than the construction of identity.

If a semiotic system designates itself as "mind" as opposed to "matter" or the other way round, this can and has throughout the tradition of Western culture led to a structure of knowledge trapped between the alternatives of idealism and materialism. The constellation of the sciences even today reflects this dichotomy. The distinction between nature and culture is a derivative of the mind/matter dichotomy. A general theory of systemic order offers the possibility of replacing this traditional dichotomy with the distinction between self-reference and non-self-reference. Displacing these distinctions frees us from a whole series of narrative interpretations that have dominated Western thought from Plato till the present day. Idealism, realism, empiricism, materialism, individualism, collectivism, and many other Western "myths" may be interpreted in new ways. Within the horizon of possibilities disclosed by traditional interpretations many battles have been fought over the nature of meaning. One influential school of thought starts from the assumption that the unity of the difference between mind and matter is being. Both mind and matter, whatever they might otherwise be at least exist. Meaning is therefore a special kind of entity or thing. There are things in the world that are self-conscious and there are other things in the world that are not. The world consists of *res cotigans* and *res extensa*.

Meaning, however, is not an attribute. It is also not a special kind of substance. No system, including a system of meaning is a "thing" in the philosophical sense of substance. Let us recall the long tradition of attempts to discover that spiritual or mental substance called the "soul" or the "mind." Discussions are still going on today about the mental attributes of the brain. If a semiotic system is not a thing at all; if there is no ghost in the machine, and no machine either, for that matter, then it is not surprising that traditional assumptions about the mind have not lead to theoretical solutions. At the level of emergent order we call meaning, there are not things, there are signs. The word "machine" and the word "mind" as well are both elements of a system of meaning. Meaning is not derived from things, but is an emergent characteristic of those elements that make up the system of meaning. If the system prefers to designate itself as spirit or as a matter, this is its business and cannot be criticized from any neutral, value-free standpoint outside the system. Outside the system of meaning there is only meaninglessness, but strictly speaking even this possibility is not available, since meaningless is one side of a difference and thus has some informational value for the system. If a meaning system is not necessarily either ghost or machine, it is also not a certain kind of organism, for example, a brain. We may have all the grey matter we want, but will have no other meaning than what we choose to call this mass of nerve cells. Meaning is not a thing or an organ, but a *level of emergent order*, namely, that level wherein things of all sorts become designated and thus are made meaningful.

A meaning system does not consist of thinking things (Descartes' *res cogitans*) or rational animals (Aristotle) or even observers (constructivism), but of "signs." Signs are the elements of the semiotic system. Things, organisms, and observers are signs within the system of meaning, no matter what they choose to call themselves and precisely because they must call themselves something if they do not wish to dissolve into meaninglessness.

These considerations bring up the very important question: What is a sign? Ferdinand de Saussure pointed out that a sign is not an entity, but a relation. Bateson assumed that meaningful information is to be relationally defined as a difference that makes a difference. To define the sign relationally as a difference instead of an identity is to emphasise that no word, symbol, gesture, or event exists by itself. A sign is not a thing, but a relation, a "value," as Saussure puts it, within a system of differential relations. To utter a single word, therefore, is to speak an entire language, as the philosopher Ludwig Wittgenstein pointed out. Language is never a mere collection of words, not even a formal differential system, despite the fact that Saussure himself and the tradition of structural linguistics that followed him supposed otherwise. To emphasize that a sign is a difference that makes a difference draws attention to the pragmatic dimension of meaning. The "pragmatic turn" in linguistics and philosophy has convincingly argued that language is best conceived of as a *chain of communications*. The elements of a semiotic system are therefore not individual signs as one might suppose, but communications, for a sign can only exist within communication. Linguistic pragmatics has made it clear that the relationing of semiotic elements is not a purely formal, logical, or syntactical matter, but a matter of social and pragmatic action. It is against this background that Luhmann could say: Society does not consist of human beings, but of communications.[7]

To say that society does not consist of human beings, but of communications, is not only provocative, it is of great heuristic value. Why is this so? Is society not responsible to the individual and to basic human rights that individuals possess? Is not the very fabric of democracy woven with free individuals who are united under the law through their own consent? A theory of meaning systems that does not base meaning on the actions of autonomous, rational subjects, but on the operations of a system that could possibly someday choose not to refer to itself as an association of autonomous rational subjects is obliged to take these questions seriously and to provide plausible answers. Despite his enormous contribution toward integrating social thought into general systems theory, Luhmann himself does not take his own statement seriously enough. Luhmann supposes that human beings are individual psychological systems that somehow "perceive," "think," and "feel" apart from or before communication. Non-communicative psychological systems

7. Luhman, *Social Systems*.

find themselves banned from society into the environment, a Hobbesian state of nature where they make up a reservoir of non-communicative, pre-linguistic meaning. Luhmann's social theory appears unable to free itself from the myth of the social contract.

Without going into the details of Luhmann's argumentation, a more coherent account would be to suppose that individuals are constructions of the semiotic system. Individuals do not observe society somehow from outside the system, that is, unless society constructs them in such a way that they are supposed to perform this function—and most societies throughout history have had no need to do this. Individualism is a product of certain basic distinctions and narrative interpretations typical of Western culture. Instead of supposing that individuals construct society, as Western social thought in the wake of Hobbes and Locke has done, it could be argued that society, understood as a semiotic system, constructs individuals. Individuals are the "persons" or "actors" that appear within society. We may speak of a meaning system as "self-organized" precisely because there is nobody—God included—outside the system who could organize it. The theory of self-organizing systems joins hands with postmodernism in that it implies a critique of individualism and the enlightenment tradition of the autonomous rational subject. Communicative actors take the place of human individuals. Actors are constructions of the system. They observe the system, paradoxically, from within. Every observation falls within the boundaries of meaning. Every observation is a communicative act, for only communication constructs sings. The system of meaning observes itself und constructs itself as communicative actor on different levels, the personal, the social, the cultural, and the ontological. This does not necessarily entail a new form of collectivism in which, once again, human freedom is compromised. Quite the contrary, this view requires merely that innovation, transformation, "critique," and impulses for social change be accounted for upon the basis of self-organization processes of identity construction within the system.

If the system of meaning consists of communications and not individuals, why do we unavoidably think of ourselves as subjects that create meaning on the basis of free will and reason? From the point of view of systems theory this question should be posed in the following way: Why are actors constructed by the system such that they perceive themselves and the world in a certain way? To pose the question in this way amounts to displacing certain basic distinctions that play in important role in structuring Western knowledge. The dichotomies of individual/collective and thought/action become dislodged from their traditional place at the roots of knowledge. Just as a general theory of systemic order displaces the dichotomy of mind/matter and the myths of idealism and materialism, the theory also allows for a reappraisal of the distinctions lying at the foundation of the conflict between individualism and collectivism. After all that has been said about the importance of self-reference for the emergence of systemic order, it will not be surprising when the theory claims that actors are constructed by the system the moment it refers its meaning giving operations to itself.

At this point the question arises: Why does meaningful self-reference produce an actor instead of something entirely different? The answer to this question lies in the answer to the question of how a system of communication regulates and steers its own communicative operations. If a meaning system may be said to consist of communications and not of individual human beings, instead of asking how individuals come together out of a "state of nature" in order to freely enter into a social contract, it must be asked how communications are selected and related to each other, and how the semiotic code controls or steers the communicative operations of the system. After meaning has emerged as a level of systemic order, it makes no sense to attempt to derive meaning from non-meaning, for there is no position outside the system from which it could be observed. There is only meaning, and beyond or outside of meaning there is nothing at all.

If we are seeking an answer to the question why we tend to think of ourselves as autonomous, rational subjects, we must begin from the premise that communicative events can only produce meaning within the boundaries of a system of meaning. Outside the system of meaning there are obviously no communicative events. This implies that one cannot "say" just anything at all. Certain things can be said and others simply don't make sense. They do not belong to the system. The difference between what belongs to the system and what does not defines the boundary between the system and the environment. In order for the system of meaning to refer its operations to itself it must know what it itself is and what the environment is. The difference between the system and the environment becomes very useful information for the system when it attempts to refer its operations to itself. This information is not only useful; it is essential, because it makes up the "identity" of the system. When the system knows what or who it is and what or who it is not, then it also knows what can or cannot be said, that is, it becomes possible to steer communicative operations within the system. We think of ourselves in a certain way because we must think of ourselves in some way or another. The way in which self-reference is interpreted is contingent. It is the difference as such that counts and not what is distinguished from what. Of course, the content of the difference does in fact make a difference, a big difference, since it determines many other basic distinctions within the system. In this sense, the interpretation of the self as autonomous rational subjectivity in modern Western culture is contingent and a peculiarity of the West, its history and unique character.

According to Luhmann, society consists of a chain of interconnected communicative operations.[8] Communications are the specific elements of the social system. If the elements of the social system are communications, then actors are only relevant to the system insofar as they communicate. Actors, it may be said, are constructed for the sake of communication. They are functional elements like the tops and legs of tables. As long as communication takes place, the requirements of system operation are fulfilled, the autopoiesis of the system continues. It is therefore possible that not only human beings may play the role of actors. Any source of communication whatever may function as an actor (see, for example, Actor Network Theory). From the point of view of general systems theory all that is required in order to construct an actor is an "identity," that is, a self-reference on the level of meaning. Groups have identities, companies have identities, institutions have identities, entire nations, supernatural beings, abstract concepts like "justice" (which "speaks" in every legal decision) and so on; all may have identities, all may be actors. Human beings, in other words, do not invent signs and use them to exchange information, as traditional views suggest, instead meaning systems construct actors (and not merely humans as every religion will testify) and use them to relate and steer communicative operations. When a world of meaning emerges out of chaos there emerges with it a myriad of actors, of self-references and sub-self-references, "who" carry and organize the communicative operations of the system. A semiotic system constructs identity upon various levels: ontologically, for the system as a whole, culturally, for a group sharing a relatively wide spectrum of frames of action, socially, for various social roles, and

---

8. Luhman, *Social Systems*.

finally personally, for individual actors human and non-human. The internal differentiation of the system, that is, the "self-organization" of *subsystemic* communication networks, institutions, groups, and finally persons occurs by means self-referential communicative operations that construct identities.

The description of meaningful self-reference constitutes the task of a unique from of cybernetics that may be termed *3rd order cybernetics*. 1st order cybernetics describes systems that operate self-referentially for the sake of the environment. 2nd order cybernetics describes systems that operate self-referentially for the sake of themselves. 3rd order cybernetics describes systems that operate self-referentially *for the sake of communication*. No theology says that God created the world for his own sake. A solipsistic God who only talks to himself (or a solipsistic observer for that matter) is absurd, as Wittgenstein's argument against the possibility of a private language (in *Philosophical Investigations*) demonstrates. A private speaker, as Wittgenstein pointed out, could not reduce complexity, maintain redundancy, and successfully negate entropy, since whatever such a speaker says is right would be right. If there is no rule to which a speaker is accountable, then there is no barrier against arbitrariness. All possibilities become equally probable, which is the very definition of chaos. It may be that all revelation is based upon God's grace, but it lies in the nature of God that there is no alternative to grace. Can there be an ungracious God? The problem of theodicy which haunts all theology shows the inherent logic of absolute symbols.

Once semiotic coding and thus communication has come into being, then absolutely everything, including communication itself can be talked about. This universal, all-encompassing and all-inclusive aspect of meaning systems is a necessary consequence of operational closure. Operational closure means that no operations of the system go outside or beyond the boundary of the system and reach, so to speak, into the environment. There are no operations of the system that do not refer to system operations. There is no communication outside the system of communication, for wherever communication occurs, it occurs as communication and thus takes place within the system. Communication leads to further communication. If it does not, then it is not communication. This explains—as every child knows—why once animals or beings from other planets begin to "speak," they become part of our world; they become, as it were, "human." A meaning system, therefore, is not only operationally closed, but also for that very reason informationally closed. All information is contained in communication. There is no information *in* the environment. Instead, there is information *about* the environment. All information, including information about the environment exists within the system. This does not mean that a meaning system has nothing outside of it, no environment. Every system, as we have seen, is constituted by a difference between itself and the environment. This general principle holds for meaning systems as well. The difference between systems of meaning and other kinds of systems regarding the relation of the system to the environment is the fact that for meaning systems *the environment is within the system.* The difference between the system and the environment is a difference that makes a difference for the system. It is, in other words, information. Paradoxically, for meaning systems, the outside is inside, since the outside is one side of a difference that makes a difference. If the environment had no meaning, it could not be talked about, nothing could be excluded from the system, there would be no language of the impossible, there would be no identifiable irrationality, barbarism, evil, abnormality and so on. If the difference between the system and the environment could not be communicated, then it could not function as a system boundary. It could not be used to steer communicative operations. The operations of the meaning system consist of communications and it is therefore by means of communication that the system must construct its boundaries, including its own constitutive difference from the environment.

The system can appear to itself in terms of the difference between subject/object, observer/observed, being/non-being, meaning/meaninglessness, culture/nature, true/false, good/evil, order/chaos, beautiful/ugly, living/dead, noble/base, sacred/profane, pure/impure, right/wrong and so on. Binary distinctions such as these are information in the technical sense of differences that make a difference. They make a difference for the system insofar as they determine the boundaries within which communication will be allowed or disallowed. They are relevant information for the system.

We may say that *what* the system at any time appears to be depends on the way in which the system *semantically* organizes itself. The structure of a system of meaning depends upon what concrete symbols are taken as starting points for what may be called operations of "de-tautologizing" and "de-paradoxing." These are the operations by which differences are introduced and information is created. By means of operations of de-tautologizing and de-paradoxing there has come into being such absolute symbols as God, the autonomous, rational Subject, the Citizen, Freedom, the Law, Tao, Emptiness, the People etc. It makes a difference for a system of meaning if it understands itself, for example, as a free association of autonomous rational subjects, or as the chosen people of God, or as Maya. What the system appears to be to itself is a matter of semantic organization. Secondly, it can be claimed that *how* the system appears, that is, the processes and operations by which the system constructs self-reference depend upon the way or ways in which the system *pragmatically* organizes itself. The pragmatic conditions of communication determine how de-tautologising and de-paradoxing operations are applied and thus how actors and what they act upon are constructed. Finally, the *media* the system uses depend upon the way or ways in which the system *syntactically* organizes itself. A system of meaning is different—as McLuhan pointed out—depending on the kinds of media communication uses.

Communication

Communication may be understood as the processing of information. This accounts for the broad range of views on the nature of communication currently available. Communication theory has emerged in recent years as a wide-ranging interdisciplinary research program focusing on the importance of electronic communication technologies and the enormous influence of mass media in today's world. Contemporary communication science can draw upon three major theoretical models of what communication is and how it functions. From the point of view of general systems theory, it can be claimed that each of these models has been developed to describe communication upon a specific level of emergent order. The classical "transmission model" of Shannon and Weaver describes a mechanical communication system, that is, the transmission of electronic signals through a channel from a sender to a receiver. The biological "adaptation model" describes how an organism constructs information internally in order to "adapt" to its environment. On the level of meaning systems we must turn to the theory of knowledge and to philosophy to find a theoretical model specific to human communication. The best candidate for this task is the "language-game model" first proposed by Ludwig Wittgenstein in *Philosophical Investigations*.

It can only cause confusion when models appropriate to the description of communication on one level of emergent order are applied to phenomena on a different level. The way in which a radio or a telephone operates cannot be used as a model for the way in which an organism interacts with its environment. The viability of an organism in a specific environment represents a unique form of interaction that must be described in its own terms. The same can be said for human communication. The way in which communication in human society functions should not be described in terms designed to describe the adaptive behavior of organisms. Nevertheless, communication science has not always clearly distinguished between different levels of emergent order and the different forms of communication that accompany them.

The model of a transmission machine was the first theory of communication to be developed and has exercised great influence upon following theories. Most attempts to understand communication today rely upon such basic concepts as "sender," "receiver," "channel," "message," "information," "noise," "redundancy," etc., all of which were defined within the context of the transmission model. The transmission model, however, is inadequate to describe communication on higher levels of emergent order. If we try to understand what happens when two people speak to each other by describing them as sender and receiver and the words they speak as messages transmitted through the air, we soon get into theoretical difficulties. How are words the same as signals? How is information selected and from what source? In what sense is speaking an encoding process and the understanding of language a decoding process? Theoretical difficulties will also arise, if we attempt to think of one person as an autopoietic, operationally and informationally closed system and the other person as the environment. How does a solipsistic subject of knowledge create and maintain redundancy, consistency, and order, when there are no external sources of information? Is the relation between two human speakers the same as the relation of an organism to its environment?

Neither the transmission model nor the adaptation model can be said to adequately describe communication on the third level of emergent order, that is, on the level of meaning systems. The transmission model describes communication as the successful transmission of signals from a sender to a receiver through a channel. Signals are not meaningful signs, but electronic impulses. Signals cause a reaction in the receiver. According to this model of communication, neither the code nor the sender and receiver are themselves part of the information that is transmitted. This is quite the opposite for communication on the level of meaning systems. A system of meaning can only achieve self-reference if it somehow identifies itself within every communication. On the level of meaning we do not know what is being said unless we also know who is speaking. The biological adaptation model is also inadequate for the purpose of describing human communication, because it cannot account for shared information. Each organism constructs its own "world" which is either viable or not viable. What appears to be an exchange of information between organisms is in reality "structural coupling"[9] or mutually adapted behavior. This view would condemn us to solipsism, as radical constructivism openly admits. Therefore, a third model, the "language-game model" based upon Wittgenstein's pragmatic description of language may be seen to be a more useful model for understanding meaning systems.

The language-game model describes communication as an inter-subjective and interactive event not limited to a single individual and not something that an individual or even a group of individuals could create. The communicative actor, what is communicated, and the understanding of what is communicated all belong to the single and unique communicative event. Human beings do not create language and meaning; rather they exist as actors within language and meaning. It is for this reason that the primary or "religious" self-reference of the system of meaning can not be left out of account, for it constitutes an essential and necessary condition of human existence.

Religion

Self-reference on the level of meaning systems implies that operational closure is achieved by distinguishing systemic identities. A system of meaning operates in order to construct meaning. The first unit of meaning is the system itself. The system must first of all make itself meaningful. Self-organization and self-reference on the level of meaning imply the construction of a meaningful "self," that is, an identity. The actors "who" make up the system must be distinguished from all that does not belong to the system. This in turn implies drawing a fundamental bound-

---

9. Maturana & Varela, *The Tree of Knowledge*.

ary between meaning and meaninglessness, order and chaos. In the case of human societies the elements of which the system consists are not, as might be expected, human individuals, but communicative actions. Communicative actions are organized such that a boundary emerges between self and other. This boundary is at the same time always a border between meaning and meaninglessness, when the self in question is the system as a whole. The uttermost boundary of the system as a whole builds a common horizon of meaning, a worldview, shared values, and a basic understanding of what is real, true, good, and beautiful. It is this "life world horizon" as it has been called by social phenomenology that makes mutual understanding, cooperative action, and a shared life within society possible. We can argue with each other, have divergent opinions, resolve conflicts, and act cooperatively within a context of antagonistic interests only because there are certain things about which we all agree. At the very least, it is necessary to admit that all parties to the discussion are human, have common concerns, and are supposed to cooperate with each other. Everything outside this uttermost boundary of shared meaning and value simply does not "fit" in the world we live in. It is marked as "impossible," "evil," "abnormal," "irrational," "barbaric," and so on. The way in which a system of meaning differentiates itself from meaninglessness, chaos, and disorder constitutes the *function of religion*. It must immediately be emphasised that the concept of "religion" introduced here is neither theological nor sociological, but a concept specific to the general theory of self-organizing systems and the theory of communication that describes how meaning systems organize themselves.

Much has been written about religion from many different points of view. There exists a long and unresolved tension in the history of Western thought between *logos* and *mythos*, reason and belief, enlightenment and superstition, science and religion. This controversy is significant not because there is a real choice to be made between religion or science, but because it points to the fact that there are functionally equivalent solutions to the problem of self-reference that compete with each other. On the level of the system as a whole there are only religious answers, but they can be very different. 3rd order cybernetics does not enter into the dispute between religion and science in order to take sides, but instead, it attempts to explain what the dispute is really about. Ideological conflict is a result of functionally equivalent communication on the religious level. As we shall see, science cannot argue against religion, only another religion can argue against religion. More importantly however, the undiminished relevance of religion in so-called secular society leads us to suspect that the religious function plays a role in every self-reference made by and within any system of meaning. Religion is important not because people believe in Gods or revelation of some kind, but because boundaries between self and other, meaning and meaninglessness must be drawn in every instance of systemic and subsystemic order. Drawing these boundaries requires a specific form of communication, a "boundary" discourse that may be characterized as religious communication. Without religious communication the difference between system and environment could not be drawn and there would be no system of meaning at all. Every identity, on whatever level it is constructed, comes into being through boundary discourse and depends in some way upon religious communication.

Anthropology, ethnology, and the history of religions have shown that there has never been a human society without religion, a worldview, or a life-world horizon of some kind. This is no accident. 3rd order cybernetics describes a system of meaning in which worldviews, ideologies, and general theories of reality play a necessary role in the construction and maintenance of systemic order. The existence and function of such symbolic worlds can be understood as the necessary self-reference of a meaning system. Not only is religion in this way understandable as a condition of the emergence of meaning and as a basis for all human communication, but "religion" also appears as a general property of semiotic order that can be analysed with concepts similar to those employed by the natural sciences for the investigation of order in nature.

This view allows for a new understanding of absolute symbols such as "God," "Freedom," "the People" and so on. Religious symbols have a specific function and have common characteristics such as inclusion-exclusion, negative and positive values, and finally a peculiar form of communication that can be analysed in terms of a unique pragmatics. The pragmatics of boundary discourse are proclamation, narrative repetition, ritual representation, temporal orientation towards founding events, and inclusion/exclusion. In general, it may be said that concepts of the absolute are nothing but the necessary self-reference of the meaning system and are constructed by communication according to specific pragmatic rules or conditions. Every meaning system, it will be seen, is based upon absolute symbols that form its primary self-reference, regardless of content and regardless or whether or not the system considers itself to be "enlightened," "rational," "scientific" or whatever. Every meaning system emerges by means of constructing absolute symbols according to specific forms of communication. The pragmatic conditions of religious communication are thus the conditions of emergent order in the human cultural realm. Knowledge of these conditions of communication within human society allows for a deeper understanding of the mechanisms of self-identification, of religious and ideological commitment, and yields as a by-product methods for inter-cultural and inter-religious understanding.

As Jürgen Habermas has pointed out, every communicative action is embedded within a horizon of commonly accepted criteria of for what is to be counted as real, true, good, and reasonable.[10] If these criteria are established on the basis of a pragmatics of boundary discourse, then they can only be transformed if there is still a higher level of discourse upon which basic distinctions can be newly interpreted. Transformation, innovation, and creation of the new are the result of a specific form of communication. Technical knowledge and logical coherence in the natural sciences, the social sciences, and also in philosophy and theology, are only meaningful to the extent that they are embedded within a communicative practice open to correction from without and possible transformation of basic concepts and methods.

Applied to human cultural systems, this means that every world view, every life-world horizon, every ideology and religion exists within a universal field of communication to which it is "ethically" responsible. Each symbolic world, whether secular or religious, whether conservative or liberal, depends for its own meaningfulness upon the recognition of other possible worlds of meaning and the openness to revision of its basic symbols and interpretations. Theories that do not acknowledge their rootedness in a discourse of disclosure are forced into oversimplifications, excessive reductions of complexity, defensive practices and finally a disfunctional and dangerous "fundamentalism". This can be observed in those ideological and religious groups that are openly fundamentalist. Transformative communication is, therefore, no less a system requirement than is conservative boundary setting. Innovation, originality, creativity, and the appearance of the new are equally necessary conditions of meaning as is the drawing of boundaries between

10. Habermas, *Theory of Communicative Action*.

what is at any time meaningful, true, good, beautiful, normal, rational, and what is considered to be the opposite of all these positive values and which therefore must be repelled and excluded from the system. Transformative communication is the function of what in contemporary Western culture has come to be identified as "art." If scientific discourse constructs meaning in the form of propositions which can be either true or false within the boundaries of commonly accepted criteria of validity and if "religious" discourse has the task of setting these boundaries, then it is "art" that transforms boundaries and communicates at the interface of exclusion and inclusion. Communicative actions on the boundary of meaning may be perceived as criminal or insane and dealt with accordingly. Those who manage to avoid this fate are neither criminals nor madmen, but artists. If there can be no science without religion, there can be no religion with art.

In summary it may be said that a general theory of self-organizing systems offers a point of view from which human society, culture and religion may be understood as systemic phenomena. The systems approach opens up new possibilities for understanding what religion is and how it functions in social communication.

7

Disruptions

Systems and Investigations

Clemens Sedmak

ABSTRACT

In my view, theological work in our times has to strive towards "local theologies" (cf. C. Sedmak, Doing Local Theology*) and a method of "theological investigations" that permits to ask the question of a "fulfilled life" (cf. C. Sedmak,* Theologie in nachtheologischer Zeit*). Is the systems theory approach compatible with these demands?*

Jesus has never developed a theological "system"; Romano Guardini has pointed out that the root of the Christian religion is a person, not a system of doctrines. Is systems theory adequate to grasp the concerns of the Christian religion?

Doing theology as if people mattered asks for a clear language and for respect for differences ("local theologies are messy"). Is systems theory capable of being developed in such a way that these requirements of a clear language and a respect for differences can be met?

PRELIMINARY REMARKS

A MAIN RESERVATION ONE might have about a way of conceptualizing human life in terms of "systems" is the fact that life is being disrupted. A system is by definition a structure, an order, a way of ordering elements. A disruption of this order represents a way of questioning the system both on the level of the particularity of the system and on the level of the very idea that a system is an adequate tool to conceptualize life. Within theology, theological systems have a long tradition. By way of theological models, academic theology tries to systematize our approaches to life.[1] Academic theology developed a culture of construing theological systems. We might ask the question whether theological systems are limited insofar as they cannot do justice to the disruptive and hence unpredictable nature of human life. An alternative to theological systems could be theological investigations. Ludwig Wittgenstein in his *Philosophical Investigations* paid tribute to the fact that language can only be reconstructed as a system if one neglects differences of the many different ways of language usage. Wittgenstein used the concept of language games as a result of his insight that we cannot make statements about language as such.[2] Philosophical investigations work on a case-by-case basis, by way of example, without the ambition to exhaust a subject manner. Philosophical investigations can come to an end at any time. One could be tempted to suggest that the adequate means to do theology is the instrument of theological investigations since 1) theology discusses human life in its fullness and depth, and 2) the concept of human life is even more complex than the concept of language.

THE CONCEPT OF DISRUPTIONS

Disruption is a process by which established points of reference for identity are being called into question. Disruptions stop the status quo. A disruption proposes a liminal state between a "not anymore" and a "not yet"—established patterns have gone; new patterns have not yet been established.[3] Disruptive processes change the coordinates of the unquestionable; of what is simply accepted. As a transitional stage a disruption invites a conflict of identities and an uncertainty that is connected with that. By way of example of marriage disruptions: "Marital disruption . . . represents a period of family conflict and uncertainty in which children may be caught in the middle—torn by their emotional

---

1. The concept of theological models has become pertinent in this context—cf. Sedmak, *Lokale Theologien*, chapter 12.

2 Wittgenstein, *Philosophical Investigations*. 65.

3. This transitional stage has been characterized by Victor Turner as a liminal stage: Turner, *The Ritual Process*; Weber, *From Limen to Border*, 525–36.

ties to both parents."[4] The concept of disruption points to the concepts of disturbance, of uncertainty, and of conflict. What does it mean to be disrupted? In his introduction to the stories from American life, Paul Auster quotes from the story sent in by an American woman whose father had left the mother so that the girl was given up for adoption. She traced her father who had married in the meantime: "We never told his wife of my existence, as I didn't really want to disrupt his life."[5] What does that mean? Disrupting his life would mean: redefining the coordinates of important/unimportant, renegotiating the space for relationships, redefining his identity in the eyes of his loved ones. A disruption of life can be understood as the drawing up of a new cosmology and a new concept of Self therein.

We could distinguish between challenges and disruptions. A challenge is an obstacle in the way that can be removed. We can continue our journey down that very road. We may have to slow down, we may even come to a stop for a while, but we do not have to change our fundamental plans. Life is full of challenges, impediments to be overcome. As long as there is a sound foundation we do have the means to cope with such challenges. Disruptions, on the other hand, are more serious a matter. Disruptions shatter the foundations of our life plans and life forms. Disruptions call for a new reading of our life; a new life-plan; a new quest for identity. Being able to be disrupted is a certain gift. The ability to cope with disruptions is a sign of life, is also a sign that someone has understood the way life works. Disruptions are invitations to begin yet again, to make a fresh start. Wisdom can be understood as the skill of dealing with disruptions. Dealing with disruptions calls for the maturity to accept things that cannot be changed and to acknowledge irreversibilities. The ability to cope with disruptions implies the flexibility of a life plan that is not densely defined. The ethics of a life-plan involves that we cannot pin down a life plan exactly to details.

The book of Job teaches many lessons on disruption.[6] Job is wrestling with his identity, with the pillars of his outlook on life and his worldview. Job is disrupted by the messages he receives through the law. He calls his whole system of orientation into question. Job is disturbed by the messages he receives. Job has to learn the lesson that a "prophetic

---

4. Bumpass, "Children's Experience," 49–65, 50.
5. Auster, ed., *True Tales*, xxviii.
6. Ticciati, *Job and the Disruption*.

language" that operates out of a certain system of judgments is limited; he has to learn yet another language, a mystical language that teaches him that God's justice cannot be measured and reconstructed in terms of human justice.[7] Job has to learn a disturbing lesson that calls for trust in God: "The most fundamental way in which God relates to creation is in such disruption and transformation. And this would mean that there are no structures of reality that are not subject to disruption."[8] The suspicion of an arbitrariness of God has to be translated into respect for the freedom of God; trust in God is the only adequate response to the disruptive nature of God's presence in our lives.

Disruption can be regarded as a key category in theology. Jesus disrupts the identity of his disciples by inviting them to take a fresh look at the categories they use, at the way of life they follow. Jesus invited them to take on a new identity with a security that is based on Jesus and on Jesus alone. "The disciples . . . are radically brought in question: not just their acts but themselves. The secure place from which they observe and judge the world is threatened, disturbed and uprooted. And they do not shield themselves from this disruption, but cling in it to Jesus, who takes them to the far side of the abyss where they regain the security they had momentarily lost."[9] A disruptive experience, caused by Jesus, is Paul's famous Damascus experience. It is a striking example of what a disruption can do to a human life.[10] Paul is on his way to Damascus to persecute Christians. He is struck down by light, not by the violence of a blow or a thunder, but by light (Acts 9:3f). Paul falls down to the ground. He receives the order to get up, only to find out that he has turned blind. He needs his companions to guide him to Damascus. He reaches the original destination of his journey, but now with another spirit, another attitude. He is blind for symbolic three days, and does not eat and does not drink (Acts 9:9). It undergoes a deep desert experience. After these three days he is healed, he has a new mission, a new purpose in life, new fundamental convictions, a new "fundamental option" (Karl Rahner). It is still his very personality that sets out for a new mission, but the context of his agency has changed.

---

7. Gutiérrez, *On Job*.
8. Ticciati, *Job and the Disruption*, 172.
9. Ibid., 43.
10. Longenecker, ed., *Road from Damascus*.

From Paul's experience we can see that a disruption can be unexpected; caused by God; an experience that makes you lose your firm foundation; you fall to the ground and find yourself in a situation of confusion and lack of orientation. You find yourself in a situation of vulnerability and dependency on others. In contexts of disruptions we are in need of companionship due to our blindness. Disruptions call to a march through the desert. It is in the desert that established life-plans are shattered and new life-plans slowly and painfully emerge. In this sense, a disruption is always the break down of a system, a system (form) of life, a mode of living, a system of fundamental convictions about one's identity, about one's life, about life and our place in the universe. Based on these ideas one could argue that theological investigations, working on a case-by case basis[11] are a more adequate means to reflect upon human life, especially since it is part of a Christian journey to allow for disruptions.

Finally, we could look into the idea that the concept of disruption plays an important role in trying to understand the concept of love. To love someone means to be willing to be disrupted by this person. If you love a person your sense of Self becomes expanded in such a way that you are willing to count as belonging to yourself the beloved and his/her life matters.[12] If something happens to your beloved you are willing to accept a change of your life on the grounds of the life change of the beloved. To love a person expresses the willingness to accept the other as a "disruptor."

CONCEPTUALIZING LIFE

The concept of disruption points to the concept of life. Life can be reconstructed as a process of disruptions. In his document *Evangelium Vitae* John Paul II characterizes life as a gift.[13] A gift, if it is accepted and received, changes the life of the recipient. It disrupts patterns of existence insofar as there are obligations attached to a gift. A gift can be understood as a cluster of commitments—the commitment to respond adequately to the giver, the commitment to use the gift wisely, the commitment to live in the light of the gift, the commitment to respond to

---

11. Sedmak, *Theologie*, chapter 3.
12. Nussbaum, *Upheaval*, part II.
13. John Paul II, *Evangelium Vitae*, 39, 49, 52.

the nature of the gift, thus incorporating the gift into one's life. Seneca in his *Epistulae Morales* talked about the importance of wisdom in order to appreciate a gift. The duly expected degree of gratefulness can only be assessed by using judgment. Reflecting upon human life can be seen as an attempt to respond to the gift of life—this provokes the question: How can we adequately conceptualize life? Unusual questions ask for unusual means: We can learn a lot about the complexity of conceptualizing life by looking at two projects—(i) George Perec's novel *La Vie* and (ii) Paul Auster's project on collecting true stories by Americans, stories about America.

Life: A User's Manual

Georges Perec in his famous novel *Life. A User's Manual* is describing the inner life of an apartment block in Paris. The building with its many flats and many interwoven life stories of its inhabitants forms a microcosm. Perec uses the image of a puzzle as a key to understanding life. The fascinating thing about puzzles is the fact that the pieces are readable only when assembled; "in isolation, a puzzle piece means nothing."[14] A piece of a puzzle does not reveal much. "You can look at a piece of a puzzle for three whole days, you can believe that you know all there is to know about its coloring and shape, and be no further on than when you started. The only thing that counts is the ability to link this piece to other pieces."[15] Once a piece has been put together with other pieces, "the piece disappears, ceases to exist as a piece."[16] It has become part of a whole, of an all-encompassing unity. If we take a puzzle to be an image of life, we can face the challenge to read each day as a piece in a whole. The pieces are, of course, not given in a predictable order. The challenge consists in reading sense into the pattern the individual pieces constitute. From a theological point of view the pieces can be regarded as an invitation to see a plan behind what is happening to us in our lives. A remark by Georges Perec in this context is theologically important: "Despite appearances, puzzling is not a solitary game: every move the puzzler makes, the puzzle-maker has made before; every piece the puzzler picks up, and picks up again, and studies and strokes, every combination he tries, and

---

14. Perec, *Life*, 189
15. Ibid.
16. Ibid.

tries a second time, every blunder and every insight, each hope and each discouragement have all been designed, calculated, and decided by the other."[17] If we take this image seriously, the disruptions in life can only be disruptions on the surface, disruptions of our construction of the whole given the various pieces of the puzzle. The plan of God can bring order even into human-made chaos; can invite a vision of meaning even in the brokenness. God's disruption of our way of reconstructing the patterns of our lives as a puzzle is an invitation to more subtlety. There is no straightforward approach to life. Perec indicates that it is "not the subject of the picture, or the painter's technique which makes a puzzle more or less difficult, but the greater or lesser subtlety of the way it has been cut."[18] A theological interpretation of this hint could consist in the invitation to look beyond the surface pattern of our lives, to look at the "depth grammar" of our lives, to look at an interpretation that gives depth and meaning to our forms of life. Working with a puzzle is hard work: "Sometimes three or four or five of the pieces would fit together with disconcerting ease; then everything would get stuck."[19] The progress on solving a puzzle is not predictable. There are "privileged instants" in putting together a puzzle,[20] moments of "grace,"[21] where the pattern of some part of the puzzle becomes instantly clear. These points are also clear when we think about our efforts to decipher our lives. The image of the puzzle for life can be considered as posing a triple challenge: a) to see life as a puzzle that is worth thinking about, as something that we have to try to understand; b) to see each day as a piece of puzzle and to regard the whole puzzle designed by God with the promise that there is a pattern behind the many various pieces; c) to see life as ever unfinished until the last day has been lived. There is no point where we could say that the puzzle of our lives is finished.

The second image of life that we can trace in Perec's novel is the image of the house, the building. Build up one's existence is like trying to erect a house, the house of one's life. The apartment block described by Georges Perec is a symbol for the cosmos, is like a polis with its many different faces. Each flat is different; each flat is made inhabitable by the

17. Ibid., 191.
18. Ibid., 190.
19. Ibid., 334.
20. Ibid., 339.
21. Ibid., 338.

details. It is the small things that make up the essence of life, not the big things. Perec talks about a person's relation to the house: "He tried to resuscitate those imperceptible details which over the course of fifty-five years had woven the life of his house and which the years had unpicked one by one."[22] The various parts and objects constitute personal identity: "The stairs for him, were, on each floor, a memory, an emotion, something ancient and impalpable, something palpitating somewhere in the guttering flame of his memory: a gesture, a noise, a flicker, a young woman singing operatic arias to her own piano accompaniment."[23] Personal identity is connected with a sense of belonging.

In his novel Perec describes the challenge of drawing up a life-plan as faced by one of his main protagonists, Mr. Bartlebooth: "Let us imagine a man whose wealth is equalled only by his indifference to what wealth generally brings, a man of exceptional arrogance who wishes to fix, to describe, and to exhaust not the whole world—merely to state such an ambition is enough to invalidate it—but a constituted fragment of the world: in the face of the inextricable incoherence of things, he will set out to execute a (necessarily limited) programme right the way through, in all its irreducible, intact entirety."[24] Bartlebooth wanted his whole life to serve a pattern, he wanted each day of his life to be centered around one single project the purpose of which consisted only in its completion. "What Bartlebooth would do would not be heroic, or spectacular; it would be something simple and discreet, difficult of course but not impossibly so, controlled from start to finish and conversely controlling every detail of the life of the man engaged upon it."[25] The idea of controlling one's life from beginning to end is threatened by disruptions that cross our plans. Disruptions can be bitter and destructive. One's existence can be deeply disrupted to the point of brokenness: "Monsieur Jérôme had not always been the bitter and burnt-out old man that he was in the last ten years of his life."[26] "He came back to Rue Simon-Crubellier in 1958 or 1959. He was unrecognisable, done in, worn out, done for. He didn't ask for his old flat back, but only a maid's room if there was one free."[27] Disruptions

22. Ibid., 61.
23. Ibid., 61f.
24. Ibid., 17.
25. Ibid., 18.
26. Ibid., 203.
27. Ibid., 204.

can mean deep humiliations, and are always connected with some kind of failing—since one particular life plan has come to an end. Disruptions always mean an experience of loss. And because of that, disruptions are always painful.

Summarizing, we could trace three main lessons on life in Perec's work: (i) A useful image for human life is the puzzle; (ii) another fruitful image of human life is the house; (iii) people develop life-plans in order to give meaning, depth and orientation to their lives.

True Tales about American Life

We can learn a lot about the disruptive nature of human life by looking at a project that was initiated by Paul Auster and a radio station. Paul Auster had asked Americans to send in true stories to be broadcast. 179 of these stories have been put together in a book by Auster.[28] This book is a living witness of the richness, width, and depth of life. A deeper look at reality is permitted if one allows space for disruptions. In his introduction Auster talks about the stories: "Of the four thousand stories I have read, most have been compelling enough to hold me until the last word. Most have been written with simple, straightforward conviction, and most have done honor to the people who sent them in. We all have inner lives. We all feel that we are part of the world and yet exiled from it."[29] What is a story that has done honor to the people who sent them in? Is it the honesty of the story? Is it the fact that the story reveals something about life that is not entirely obvious? Is it some insight about life the story may contain, some unexpected (in this sense: non-trivial) truth? Is it the challenge of disruption?

> I learned that I am not alone in my belief that the more we understand of the world, the more elusive and confounding the world becomes. As one early contributor so eloquently put it, "I am left without an adequate definition of reality." If you aren't certain about things, if your mind is still open enough to question what you are seeing, you tend to look at the world with great care, and out of that watchfulness comes the possibility of seeing something that no one else has seen before. You have to be willing to

---

28. Paul Auster, ed., *True Tales of American Life*.
29. Ibid., xvi.

admit that you don't have all the answers. If you think you do, you will never have anything important to say."[30]

This seems to be an indication that disruption is a key element in the perception of the world. If you want to draw a deep picture of life, or a picture of life with its depth dimension, you have to allow for the unexpected, the disruptive, the moment of suffering perhaps. The stories teach us a lot about human life—"incredible plots, unlikely turns, events that refuse to obey the laws of common sense."[31]

If we look at these stories they do reveal general insights about human life. I will reconstruct four lessons about life with special reference to life's disruptive nature:

(1) Disruptions at a certain moment lose their urgency in the light of the larger context of life. It is a sign of wisdom to be able to place experiences and decisions against the background of the whole of life.[32] A boy lost his sister's parakeet because of carelessness. "Kathy managed to forgive me. With fake optimism, she even tried to reassure me that Perky would find a new home. But I was far too canny to believe that such a thing was possible. I was inconsolable. Time passed. Eventually, my great remorse took a modest place among the larger things of life, and we all grew up."[33] Disruptions like loss of something precious confront us with irreversibilities of the kind that do not allow us to say: All will be the way it used to be. A language of "all will be well" in this situation calls for an attitude that things will be different, yet well. It is a sign of wisdom to place things into a larger context. It also on the basis of this perspective of a larger context that disruptions and fragility can be seen as aspects of being truly human:

> I realize that I cherished my father for his strengths. He had a spirited investment in taking chances and a bottomless reservoir of optimism ... My father was always eager to try new things, to bring fun changes to our lives. Sometimes one or the other of us was reluctant or afraid, but he had a way of encouraging us to take a chance ... Now that I look back on my father, under the focus of forty-four-year-old eyes, I know that what I loved most

---

30. Ibid., xvii.
31. Ibid.
32. Cf. H. Krings, Sapientis est ordinare; Assmann, *Was ist Weisheit?*
33. Auster, *True Tales of American Life*, 30.

about him was his fragility. And because I sensed this, I developed a desire to protect him. I think everyone in my family felt the same. We were in awe of his exuberance but cradled fear for him, too. Maybe he carried so much sense of promise that we realized how hard it would be on all of us to see him disappointed, disillusioned, or hurt."[34]

Jean Vanier in his remarkable insight into human nature discusses the importance of being vulnerable in order to do justice to the human condition.[35] Vulnerability is an important aspect of being human. In the larger context of life we discover general "laws of life." It is a general law that, at some point, parents become weaker and children become stronger. There is such a thing as an order of generations.[36] This point was illustrated in one of the experiences: "At that moment, with my dad's tears falling on my shoulder, I felt like I was his father and he was my son, and in the solace of my arms he discovered the safety I had once sought in his."[37]

(2) Some disruptions are powerful and leave wounds that cannot be healed, open up books that cannot be closed.[38] There are experiences that cannot be healed; there is injustice that cannot be put to right. An American confesses of an injustice done to a Japanese prisoner whom he never saw again: "Now, fifty years later . . . I wish I could find the man, so that I could apologize to him."[39] What does it mean to live with this open story that cannot be closed? Some disruptions leave us with a concept of life falling apart:

> Things began falling apart for us in the summer of 1930. That's when my father refused to take a cut in pay and wound up losing his job. He spent a long time looking for something else, but he couldn't find work, not even for lower wages than what he had already turned down. Finally he settled into a chair with his *Argosy* magazine, and my mother began to nag and fuss. Eventually we lost our house . . . I remember a dream I had about finding jewels

---

34. Ibid., 34f.
35. Vanier, *Becoming Human*.
36. Robert Nozick who had gone on this search of such "laws of life" made this point well—cf. Nozick, *Examined Life*, 28ff.
37. Auster, *True Tales of American Life*, 147.
38. Schimmel, *Wounds not Healed*.
39. Auster, *True Tales of American Life*, 74.

fort hem, but when I put my hand in my pocket, all I found was a hole. I woke up crying. I was six years old."[40]

Basic trust in life was taken away from that child. Some situations are beyond repair. Another child was confronted with his father's being laid off. After a long time of searching for an adequate job "my father lost his spirit, any hope that his last working years would be of real value."[41] Hence, it has to be part of any reflection on life to deal with things that cannot be undone, to deal with irreversible decisions and mistakes beyond repair. It is systematically misleading in the construction of ethics to deal with ethics as if we were confronted with a white sheet of paper, a tabula rasa.

(3) Life teaches us lessons through significant moments, significant experiences. "Once, when I fainted in Oklahoma, my mother waded through poison ivy to get to a stream and dip a cloth in the water for me. By the time we were in Texas, her legs were so swollen that we had to stay in Dallas until she could walk again."[42] The same story recounts the unexpected encounter with a charitable and generous person: After this encounter "I had learned all I would ever need to know about charity, faith, trust, and love."[43] Disruptions direct our lives in a new, unexpected direction. Looking back on our lives we can identify significant moments that changed our life sustainably—in many cases we have to ask ourselves what would have happened if ... : "What would have happened if I had slept in? What gave me the itch to go to the Rose Bowl on that particular day? What if I hadn't turned that last corner, choosing instead to leave and rest my aching feet?"[44] This experience is not uncommon. Looking back on the life of a man who never found his place in life a person writes about a piece of work delivered at school: "His essay was not only a show of adolescent anger, but a turning point in his life: a turn toward a darker, difficult future which has never, to this day, sorted itself out."[45] However, it is important to realize that we can create significant moments by our decisions. Human beings are able to redefine their lives

40. Ibid., 117.
41. Ibid., 102.
42. Ibid., 118.
43. Ibid., 119.
44. Ibid., 48.
45. Ibid., 135.

at any given moment.⁴⁶ Even though we might not be able to change facts, we are able to change attitudes. That is why attitudes is an ethically relevant topic in the search for an "ethics of life." Recapturing a sense of awe and wonder of being alive: "In some subtle way I am changed. I can feel the sun on my skin, see my dog's face, and hear the birds singing. In a world where life is sometimes mundane, repetitive, and often cruel, I am filled with wonder."⁴⁷

(4) Life is about finding one's place in life; a place where we can grow; a place where we can be ourselves; a place where we can become the persons we are called to be, given our talents, gifts, limitations. We try to make a home in our life circumstances. Part of "being at home" is people, and part of being at home is objects. A person talks about saying farewell to a car: "After all, we'd had some good times in that car. Good-bye is good-bye, even to an inanimate object."⁴⁸ A life place is constituted by commitments and important commitments are also emotional bonds that tie us to objects with a history.⁴⁹ One's personal history is significant in the understanding of human life. Life cannot be reconstructed in terms of general time-keeping. There is a personal chronology attached to life. This is part of having a particular life, one's individual life. One cannot live the life of another person: "It was the year my mother stopped drinking, so it was two years after a careless driver killed my sister in a crosswalk, one year after my father died of a massive coronary on the front stairs, eight months before my brother Ronnie died of AIDS, and six months before he revealed his situation."⁵⁰ We must not forget that the concepts we use, the things we deal with, the convictions we hold are linked to a personal history. The search for one's place in life can be painful. We encounter sentences like: "I worked at a job I didn't like very much. After a while, it made me not like myself very much";⁵¹ "I was being strangled by the wrong life";⁵² "For half an hour I crouched in

---

46. This is an important thesis in Avishai Margalit's book, *The Decent Society* and, of course, part of Viktor Frankl's approach to "responding to life."

47. Auster, *True Tales of American Life*, 80.

48. Ibid., 61.

49. This is how to characterize not only but especially moral emotions—by the characteristic history; cf. Wollheim, *On the Emotions*, chapter 3.

50. Auster, *True Tales of American Life*, 107.

51. Ibid., 75.

52. Ibid., 74.

the middle of the river wondering what I was going to do with my life."[53] On the way to find our place in life we make wrong decisions, blurred by various factors, especially greed. A person talking about a bad decision that caused him a terrible year admits: "The contract and the money clouded my judgment."[54]

These lessons make clear that life can be conceptualized using the concept of disruptions. Disruptions at a certain moment lose their urgency in the light of the larger context of life. Some disruptions are powerful and leave wounds that cannot be healed, open up books that cannot be closed and change life sustainably. Life is shaped by disruptive moments, by significant situations that give life a profile and a direction. Life is about finding one's place in life, a place where we can grow and where we can deal with disruptions.

THE ENTERPRISE OF THEOLOGY

Theology as a systematical and methodical discipline calls for the construction of systematic theological models. Models are speculative instruments and heuristic constructions that systematize relevant aspects of a given context.[55] In this sense there is an indispensable "system dimension" in the enterprise of doing theology. This systematicity must not, however, lead so far as to disallow the possibility of disruption. Popper's concept of falsification has prominently shown that a theory in order to be open to rational criticism has to have entry points where this theory is challenged. For good reasons, Popper's idea of falsification cannot be applied to religions.[56] It is also highly questionable whether the concept of falsification is useful for philosophical and theological contexts given the categorical character of these disciplines.[57] It seems more plausible to use the concept of "disruptibility" for the context of theology. Theological models "provide ways through which one knows reality in all its richness and complexity. Models provide a knowledge that is always partial and inadequate, but never false or merely subjective."[58] Hence, models are

53. Ibid., 77.
54. Ibid., 43.
55. Wolters, "Model," 911–12.
56. Sedmak, "Die Frage," 321–51.
57. Muck, *Rationalität*; also St. Körner, *Categorial Frameworks*.
58. St. Bevans, *Models of Contextual Theology*, 24.

asked to reflect the complexity of the context of reference. If theology is about life and if life is beyond indisruptable conceptualization it makes sense to think through the idea that a decent theological model is open to disruptions. In analogy to the theory of defeaters we could develop a theory of "disruptors." A disruptor is an incidence that sustainably changes the claim of a model. A decent theological model can be challenged in its claims by situations and argumentative structures.

This point can be made even stronger by looking at the specific nature of theology. Theology talks about God in the mode of a love relationship. This love relationship can be expressed in terms of disruptions. C. S. Lewis discovered the iconoclastic nature of God in his remarkable reflections *A Grief Observed*. After having lost his beloved wife C. S. Lewis comes after many days of struggling and mourning and wrestling with God to the point where he accepts the disruptive nature of God. Theological models that want to do justice to this "Magis" of God beyond theological conceptual capabilities[59] no theological model can afford to regard, e.g., the theodicy as a problem that can be solved. It remains a mystery because of its very nature of disrupting human cognitive efforts.[60] We can learn a lot from Jewish philosophy about the disruptive nature of God who engages in a dialogue with humanity.[61] The true nature of a dialogue comprises a moment of surprise and unexpected conversational moves. Theology without this dialogical moment (allowing God to speak and being listeners of the Word) turns into philosophy of religion. And it is this dialogical character—with all the implications as pointed out by Martin Buber—that makes theology a discipline that deals with life, human life in the light of the living God. Blaise Pascal's *Thoughts* with his plea for an alternative "logic of the heart" can be read as a reminder that systems are limited and the limits are set by the disruptive nature of God and human life.

There is nothing wrong with theological systems as long as they are disruptable. Any serious effort to do theology after Auschwitz has accepted that Auschwitz is a disruptor of philosophical and theological

59. IV Council of Lateran, 806.

60. The Catholic Church has conceded that the question of reconciling God and human suffering is a mystery (Catechism of the Catholic Church 309–14). The unsolved challenge remains also in the field of philosophy; a point that has been convincingly been shown by Susan Neiman, *Evil in Modern Thought*. In this sense, human suffering has to be accepted as a disruptor that cannot be destroyed with conceptual means.

61. Leaman, *Evil and Suffering*; Plaskow, "Facing the Ambiguity," 510–12.

thinking.⁶² It is part of theological reflection upon life to look out for disruptors. It has been discussed whether 9/11 can or should be seen as such a disrupting force changing the claims of models and theories and calling for new models and theories. These are genuine theological reflections. Theological investigations react to challenges. They are called for when something of a disruptive nature occurs. Theological investigations can be put together in form of a system. But the system has to be construed in such a way that it can be shattered by theological investigations. In this sense Pascal's *Thoughts*, Wittgensteins *Philosophical Investigations* and Rahner's *Theological Sketches* are examples of a way of thinking that does justice to the disruptive nature of life.

---

62. Cohen, *Tremendum*; Fackenheim, *Jewish Bible*.

8

Portal, System, and Sacred Order

America

PETER MURPHY

ABSTRACT

Around World War I, American universities became involved in promoting academic programs and scholarship on the theme of civilization. Partly this was the result of America's entry into the European war—itself a function of America's tentative emergence as a world power. Reflection on the nature of civilization was a search for the intellectual frame for America's world role. Partly it was also a function of America's emergence as a force in the arts and sciences. Civilization was a shorthand term for ebullient intellectual creation.

The other significant cause of the swell of interest in the theme of civilization was the impact of continuing large-scale immigration to the United States of Catholic, Orthodox, and Jewish populations from Southern and Eastern Europe. This immigration of poor, labouring and non-Protestant groups was a subject of both social and intellectual anxiety. On the intellectual front, it provoked questions about the role and place of the non-Protestant mind in the predominately Protestant intellectual landscape of America.

In response to all of the above, American academics and intellectuals created a new model of civilization: "Western Civilization." This model was inclusive of the contribution of Greeks, Romans, Jews, Catholics,

Protestants, Romantics, and Pragmatists. It eschewed both the developmental focus of Scottish Enlightenment and Marxist universal history, and the pessimism of Spengler's philosophical history. It was also more sceptical of the non-Western contribution to intellectual creation than Toynbee.

The current essay does a number of things. Firstly, it looks at the role of Catholic intellectual milieus—Neo-Aristotlean and Neo-Thomist currents—in creating the climate for and the model of "Western Civilization." Secondly, the essay explores the strengths and weaknesses of that model. Thirdly, the current essay discusses the way in which a number of the key ideas of Ludwig von Bertalanffy provide clues to correcting and overcoming the frailties of the "Western Civilization" model without denying its strengths. In doing so, the essay returns to the themes of Bertalanffy's "system theoretical concept of history" in General System Theory.

The essay adapts Bertalanffy's notions of open systems and self-organizing order ("organization") to show how civilizations with high levels of artistic and scientific formation rely on autopoietic ecumenes in order to emerge and sustain themselves. The essay shows how the model of the autopoietic ecumene provides a more precise account of civilization without comprising the insights of Neo-Aristotlean universal history. The revisionist model as applied to America shows how autopoietic ecumenes conditioned the very rise of neoAristotlean and Neo-Thomist thought in North America.

CIVILIZATION AND THE WEST

AT THE BEGINNING OF the twentieth century, the United States emerged as a world power. One of the steps that it took, to prepare itself for this role, was to embrace the idea of Western Civilization.[1] For a time, it seemed as if this might serve as a postulate for America's new global ambitions and responsibilities. Western Civilization was a grand, but vague, narrative—a philosophy of history of Greeks, Romans, Jews,

---

1. A July 2004 search on the English language title keyword phrase "Western Civilization" in Harvard University's Hollis library catalogue shows something of the history of the phrase. The first appearance of "Western Civilization" as a book title was in 1868 (*The Influences of Western Civilization in China*). This was followed by a small handful of titles on Western influence in East Asia, and then in 1898 the pace picks up (beginning with Cunningham's *An Essay on Western Civilization in its Economic Aspects*), and thereafter on average there is a title per year till 1947. This is then followed by an explosion of titles—e.g., six in 1948, nine in 1951, six in 1960, eight in 1964, and onwards at this rate till 2004.

Catholics, Protestants, Romantics, and Pragmatists. It drew widely on the seedbed cultures of the Eastern Mediterranean but owed much to the legacy of Charlemagne and his creation of a territorial European empire.[2] From its Foundation, America was skeptical of Europe and its narratives. Europe was synonymous with persecution, despotism and terror. The notion of Western Civilization, though, convinced enough Americans that Europe was worth saving. It justified American entry, at enormous cost, into the First and Second World Wars.[3]

The idea of Western Civilization found its eventual epitome in the Chicago-based publishing program organized by Robert Hutchins—Encyclopaedia Britannica's Great Books series (1952–)—and in Western Civilization teaching programs in American higher education.[4] Its influence was greatest during the First World War and the early years of the Cold War with the Soviet Union. However, when America got into trouble fighting an aimless war in Vietnam, the taste for Western Civilization soured. In the late 1960s the American intelligentsia turned decisively against the notion.[5] The West became identified with decline, war, slavery, empire, exploitative globalization, suicidal pathology, ecological mayhem, spiritual crisis, techno-scientific domination, and much

---

2. The empire of Charlemagne (742–814) incorporated what today are Switzerland, France, Belgium, and the Netherlands, plus half of Italy and Germany, and parts of Austria and Spain.

3. It was also the basis for a program of civic education aimed at migrants from the margins of Europe, from its Eastern and Southern rims, pouring into New York and other great American cities. See, Carnochan, *Battleground of the Curriculum*. The migrants had allegiances that were either uncertain or simply threatening to America's Anglo-Protestant majority. Western Civilization was an inclusive enough concept to fudge the distinctions between Protestant, Catholic, Jewish, and even (at a stretch) Orthodox heritages. It also instinctively bridged between what had been the historic predominance of migration to America from the Northwest European littoral (Anglo, Scottish, Irish, Dutch, Nordic) periphery and the swell of migration from Continental Europe as well as from Europe's Southern littoral.

4. Hutchins, *Higher Learning*. From 1943 until he retired in 1974 Hutchins was chairman of the Board of Editors of Encyclopædia Britannica. He served as editor in chief of the 54-volume Great Books of the Western World, published from 1952.

5. For a critical though wry account of the salvos by the American higher educated against "Western Civ," see, Allan Bloom, *Closing of the American Mind*. Of course, the irony was that Bloom himself was by no means an enthusiast for much of what Europe in the modern age had produced—i.e., nihilism. Bloom saw himself as a defender of American common sense against the philosophies of Heidegger and Nietzsche and Rousseau in particular. This did not stop him from loving the gourmand life of Paris. For a portrait of the sybaritic Bloom in Paris, see Bellow, *Ravelstein*.

else that was untoward. More curiously, the West was now identified with America. In a head-spinning turn, American intellectuals embraced European philosophies and engaged in a cultural war against "Western" (read: American) thinking, while Europe, the historic locus of the West, was excused its culpable history of tyranny.

Despite the Eurocentric despair of the American intelligentsia, the West remained throughout all of this a buoyant geo-political symbol and reality. It represented the return of Europe from the civilizational catastrophe of totalitarianism. For a time the idea of a Western civilization and the policymaker's vision of the geopolitical anti-totalitarian West of the post-1945 era coincided. By the 1970s anti-Americanism resurfaced in countries like Germany and began to drive the twin geo-political and civilizational conceptions of the West apart. But their temporary coincidence had already made a huge difference. After the Anglo-American defeat of Nazi Germany, they provided the conceptual umbrella under which Charlemagne's Empire was rebuilt as the European Union—eliminating the fratricidal relations between France and Germany, and overcoming the temptations of totalitarianism. When, after prolonged entropy, the Soviet Union collapsed in 1989–1991, it became possible to incorporate a range of states on the Eastern margins of Europe into this project, exceeding even Charlemagne's reach.

America, post-1945, was the guarantor of a re-born European civilization. Extraordinary numbers of Americans lost their lives in the First and Second World Wars in order to rescue Europe from fratricidal and totalitarian misery—400,000 in WWII alone.[6] In the post-war era, the primary burden for the defense of Europe was carried by the United States. The conjugation of American and Europe under the rubric of the West was a logical reflection of close trading, investment, and military ties after 1945. Yet, for all of this, America and Europe were an odd pairing. America may have been *for* Europe but it was not *of* Europe. Nothing in Charlemagne's imagination would have prepared him for the New World, and much about the Old European World worried Americans, not least the black holes into which so much of Europe had fallen in the course of the twentieth century. America may have rescued Western civilization from totalitarian barbarism, but a persistent question remained whether America belonged to the West or whether it was simply

---

6. A useful point of comparison is the 300,000 Union soldiers who died in the American Civil War.

the guarantor of the West? No matter how often policy makers might invoke the geo-political imperatives of the West, and no matter how real these were, many, if not most, Americans (including many, if not most, policy makers) saw the United States as an exceptional society. It was an exception to the corruptions of the world at large and in particular to the corruptions of Old Europe ("Core Europe"), an indispensable part of any definition of the West.

Throughout American history, Europe has always represented vice counter-posed to American virtue: the Old World to the New World; Old England to New England; Egyptian bondage to the Promised Land; despotism to liberty; indentured serfdom to yeoman farming; primogeniture to free-holding; manufacturing misery to agrarian Eden; big cities to small towns; decadence to purity; paternalism to self-help; cynicism to innocence; status to achievement; inertial crowds to expanding frontier; patrician capitalism to progressive industrialism; global markets to domestic markets; national protection to free trade; free trade to tariff barriers; empire to democracy; overseas empire to continental empire; landed empire to maritime empire; Catholicism to Protestantism; wage slavery to property owning; dogmatic truth to libertarian opinion; commercial speculation to producer rationality; laissez-faire to scientific management; trusts to laissez-faire; state socialism to welfare liberalism; bureaucratic collectivism to individual freedom; dictatorship to law; family capitalism to intellectual capitalism; the French Revolution to the American Revolution; terror to elections; early retirement to hard work—in short, entropy to high-energy order.

Sometimes American impatience with Europe expressed itself through isolationism and inward-looking nationalism. The notion that the U.S. is a spearhead of the West runs against the grain of much conservative and liberal sentiment in America.[7] American ideologists of widely varying stripes think that American virtue flourishes best in isolation, and that the West is tainted either with an unacceptable history of aggression and domination or else with an equally unacceptable history of secularization and nihilism.[8] At the same time, exceptional-

---

7. In 1821, John Quincy Adams put it in these terms: "America, with the same voice which spoke herself into existence as a nation, proclaimed to mankind the inextinguishable rights of human nature, and the only lawful foundations of government . . . But she goes not abroad, in search of monsters to destroy." See Weeks, *Building the Continental Empire*, 62.

8. Charles Beard in the 1920s developed the mature form of the national liberal view, which painted America overseas as an inveterate imperialist. William Appleman

ism sometimes is expressed through outward-looking interventionism. Late twentieth-century neo-conservatives saw America as a force for the global spread of democracy. They rejected both liberal criticism of America as an unconscionable aggressor-dominator and the national-conservative preference for a less perforated society. Neo-conservative America was internationalist. Yet it still stood apart from most other nations. Notably it took issue with Europe's reluctance to confront modern vertiginous despotism.

For all of the unquestionable uniqueness of American society ("only in America"), its exceptionalism paradoxically is not an exception. Indeed this exceptionalism increasingly has channelled itself through America's strong elective affinities and alliances with other settler societies and its "special relationship" with the United Kingdom. It is notable that the UK produced most of the modern settler societies—not least the United States itself. In interesting ways, America's ambivalence towards Europe echoes the United Kingdom's own ambivalence to the "European idea" of Konrad Adenauer, Robert Schuman, and Jean Monnet—architects of post-1945 European integration. Winston Churchill's careful depiction of Britain as both *inside and outside* of Europe epitomises this. In his famous speech in Zurich in 1946 Churchill urged the construction of a "United States of Europe." Yet he did not see the United Kingdom as part of that project but rather its "friend and sponsor."[9] British exceptionalism in relation to Europe is not all that different from the exceptionalism of the United States. It ultimately rests on the peculiar social physics of Britain as a society that is the product of waves of invasion, conquest, and migration—from the Romans through the Danes, Saxons, Normans, Huguenots, Dutch, Jews, Caribbean Islanders, Africans, Pakistanis,

---

Williams recapitulated this view in a vaguely Marxist version. See Williams, *Tragedy of American Diplomacy*; Williams, *Roots of the Modern American Empire*. More recently it was re-done in an institutionalist guise by Chalmers Johnson. See Johnson, *Sorrows of Empire*. On Beard's work, see Noble, *End of American History*; Benson, *Turner and Beard*. On American exceptionalism, see Lipset, *American Exceptionalism*; Glaser and Wellenreuther, *Bridging the Atlantic*; Hietala, *Manifest Design*; Madsen, *American Exceptionalism*; Gutfeld, *American Exceptionalism*.

9. Winston Churchill, "The Tragedy of Europe." Britain's membership of the EEC, EC, and EU made no difference to the template that Churchill envisaged—of a Britain that had close economic and military relations with Europe but a strategic relationship based on common values with the United States. On his subtle mix, see Jenkins, *Churchill*.

Indians, not to mention the complex patterns of internal conquest and migration and circulation between the English, Scots, Welsh, and Irish.[10]

American exceptionalism is the greatest expression of the exceptionalism of the settler society cohort. This cluster of societies has a discrete character, and one that cuts across the East/West distinction. Indeed, among the most successful modern settler societies are Taiwan, Hong Kong, and Singapore. While ultra-Modern and often even quasi-Anglophone, the East Asian settler states are also indelibly Chinese. The exceptionalism of the settler societies makes sense of some things that otherwise are hard to explain—for example the very close ties of the United States to settler states like Israel, Taiwan, and Australia in spite of their relatively small size.[11] Thus, while American policy makers spend vast sums on the defence of Europe, they spend vastly more per capita on defending the Israeli settler state.[12]

The West is a fraught concept. Historically it derives from the geopolitical division of the late Roman Empire into eastern and western zones. The West was also a symbol of civilized order during the anarchy and entropy of Europe's Dark Age. Order meant in effect "a new Rome." Unsurprisingly the architect of the first Europe, Charlemagne, went to Rome in order to be crowned Holy Roman Emperor by the Pope in 800. Yet, while the constellations of Rome and Europe shared certain characteristics, Europe was no renaissance of Rome. Both had world ambitions. Both looked beyond the conventional social scale and geographic reach of commands and rules. Both had a strong impetus toward social self-organization. But the European model of autopoiesis was that of the Creator God. The Classical model in contrast invoked Nature's God. The classical image of Nature's God extended back as far as the pre-Socratic Greek idea of *phusis*. Classical Nature was a lively universe of forms built out of symmetries, proportions, scales and rhythms, and animated by a world spirit or *pneuma*. The European model, in contrast, drew heavily on the idea of genesis. Divine "origin" rather than sacred "order" was the most important characteristic of European or Western Nature.

---

10. This was already observed by Emerson. See his "English Traits" (1856) in Bode & Cowley, *The Portable Emerson*, 423.

11. This echoes British ties to Singapore, Hong Kong, and Australia.

12. Between 1976 and 1985, a quarter of all U.S. economic and military aid went to Israel. This was equivalent to 13 percent of the Israeli gross national income. See Ferguson, *Colossus*, 113.

The implications of this have cast a long shadow. Self-organizing societies are mediated in one of two ways—either by orderly morphological patterns or by the explosive upsurge of genesis out of nothing and the return to incommensurable (dynastic, biblical, racial, national) sources. Three clusters of self-organizing societies have appeared in history to date—the ancient Greco-Roman, the Eurocentric Western, and the Anglo-American-Settler kinds.[13] Each has relied on either pattern or genesis thinking. For all of its pyrotechnic modernity, America turned out to be much closer in spirit to the autopoietics of Greco-Roman antiquity than Europe. America today harbors many weird ideologies—from Old Testament moralism to Hollywood liberalism. Bewildering varieties of apocalyptic and technocratic, libertarian and communitarian world views share the same public space. They can do so because their influence is small compared with the automata-like workings of constitutional balance and public order instituted by the America's Deist Founders. What the Founders were after was a mimesis of Nature's God. Both Thomas Jefferson and James Madison pointedly invoked this phrase, as did Tom Paine. The European story was quite different. Whereas creationism became a durable but risible oddity in America, romanticism ended up as Europe's all-consuming ideology and the crucible of race empires, totalitarianism, fascism, and terrorism.[14] While Europe borrowed ideas of nature and form from the Classical world, from the ninth century onwards the Eurocentric West developed its own Faustian (Romanesque-Gothic-Baroque-Romantic) culture. In this culture, form gave way to infinity, and the yearning for infinity was fueled by a desire for negation. The Faustian world pioneered the equation of "being" with "nothingness." Its eros lusted after the null, the void, the zero—the oblivion (*lethe*) and the *nihil* (Nothing) that lay beyond the boundary of existence.[15]

If plastic figures and well-bounded bodies defined Classical culture, a taste for the immeasurable and illimitable characterized the Faustian culture of Europe. While there are major overlaps between the Settler

---

13. There is no reason that other examples of such order cannot and will not appear in history. But such breakthroughs are rare.

14. Murphy and Roberts, *Dialectic of Romanticism*.

15. On the appearance of the culture of death in the late Renaissance and early Barqoue, see Carroll, *Wreck of Western Culture*. On the Baroque as an entropic crisis-obsessed culture, see Peter Murphy, *Civic Justice*, 193–220; on the lethal trajectory of Central European Romanticism, see Murphy and Roberts, *Dialectic of Romanticism*.

Society Cohort and the Eurocentric West, there is also an important difference between them. The former have been peculiarly resistant to the Faustian component of the Western makeup. Faustian impulses have influenced the settler societies. Their military, businessmen, and intellectuals have been routinely touched by Faustian urges. They all have experienced bloody episodes of excess but nothing at all to compare with the awful thanatocratic history that extends from the holy war of the Crusades to the pan-national totalitarianism of Hitler. Settler societies, when tempted by extremism, have usually pulled back from the brink. They have managed for the most part to deflate the lure of Faustian storms with a love of grace, balance and equilibrium.[16] They have been deft in deflating "storm and stress"—by turning Romantic nihilism into charming landscapes or Baroque grandiosity into mild rituals of state. In this way, they have avoided the propensity of Faustian culture to turn self-regulating order into self-annihilating disorder—and high energy into obsessive death seeking.[17]

FREEDOM AND ORDER

If America and Europe are so different, why did the narrative of Western Civilization become popular in the American imagination at the time of the First World War and after the Second World War? One reason is that it addressed important questions about the nature of order and anarchy. American political thought traditionally has focused on questions of contingency: freedom, change, choice, hope, and opportunity. It is difficult to over-estimate just how strongly these resonate in the American imagination. But there are also fundamental aspects of the human condition that these themes do not address. Self-regulating order is one of these. The idea of autopoietic civilization explains how order is possible in societies that exhibit high levels of social or personal contingency.

In certain respects it does not matter whether we are talking about the Greco-Roman world, the Eurocentric West, or America and the Settler Cohort, the basic condition of order in a free society is the same. Freedom requires form. Without it, autopoietic (autonomous) societ-

---

16. The same cannot be said for Japan in the 1930s, or Islamic countries like Saudi Arabia and Pakistan in the 2000s, where the mix of local religion and imported European nihilism created deadly political movements. In both cases, the only resistance to this was Anglo-American and from the settler societies.

17. A classic example of the later was Spain's conquest of the Americas.

ies collapse into chaos or nihilism. Contingency without order; choice without nature; opportunity without social physics; and freedom without design—all of these turn human energy into a shapeless waste. Yet order is also often confused with dictatorship, hierarchy, or law-fixation. Such false order strangles human energies. It is as debilitating as nihilism but in the other direction. Overall, too little structure dissipates energy while too much structure squeezes the life out of energy. Either way the result is depression and entropy. Avoidance of entropy requires open social systems. These allow the import and export of social energies across system boundaries. But the freedom of traffic between systems becomes destructive unless these boundaries can also be maintained, ensuring systems have an identity. The "form dimension" of social organization operates at a pre-linguistic level. In contrast the "freedom dimension" of social systems is overtly, and often loudly, linguistic in nature. It is articulated through explicit ideologies and claims to rights.

Americans have been very fertile in devising political ideologies: progressivism, populism, welfare-state liberalism, neo-conservatism, and post-modern liberalism among them. Most of these ideologies had their origins in American borrowings from European natural rights ideas. The dominant Lockean strand in American political thinking equates nature with rights. This applies even to ideologies that refuse the title liberal. In America, "liberal" has narrow as well as broad connotations. Voters sub-divide into the Big Three cohorts of liberals, moderates, and conservatives—about a third each of the voting populace. Yet almost everyone uses the "language of rights." Thus, American romanticism presents as antinomian liberalism or as libertarian conservatism while neo-conservatives prefer Anglo-Scottish to French-Continental Enlightenment.[18] In the end most American ideologies are variations of the "grand American liberal tradition" as Allan Bloom called it.[19] Only genuine reactionaries (racists, xenophobes, misogynists) exempt themselves from this.

The equation of nature with rights, however, presents a problem. It powerfully unites, but also confusingly conflates, pre-linguistic nature with the declaration of rights. This means that natural rights ideologies have only a limited efficacy. This becomes clearer when we begin to consider society from the standpoint not of "rights and liberties" but of

---

18. Kristol, *Reflections of a Neo-Conservative*; Novak, *The Spirit of Democratic Capitalism*.

19. Bloom, *The Closing of the American Mind*, 334.

meaningful "order."[20] To be clear—let us assume natural rights as a given. Let us not enter into a quarrel with natural rights, as reactionaries do, but then let us also go one step further and ask: how does a society, with widely-dispersed "rights and liberties," create order for itself? In other words, how does this society solve what Talcott Parsons (1902-1979) called "the problem of order"?[21] How does it create a sacred canopy for itself—an ordering of things that gives coherence, pattern and meaning to human deeds?[22]

Historically, American political ideologies had much to say about freedom but shied away from a protracted discussion of how social order is possible. And yet a deep impatience with the recurrent disorders of Europe underlay American thinking in general. Until the beginning of the twentieth century, this gap could be papered over by saying that the disorders of Europe were the consequence of despotism and that freedom was the antidote to disorder. By that time, though, there had begun to appear a phenomenon that the category of despotism in the classic sense did not fully explain. By 1918 in Europe, the Balkans, and European-controlled Africa, a series of holocausts had taken place. These went far beyond the bounds of normal political violence. In these holocausts, death had emerged as the end of politics, rather than just one of its means. Death's rationale was no longer fear or submission, victory or conquest. Death's justification instead had become death itself.

The idea of civilization was a retort to this. "Western civilization" promised to unite Americans and Europeans against "thanaticism." Yet European culture was a primary source of this lethal trend. Representatives of Europe's civilization openly offered the gift of death to humankind. The German philosopher Martin Heidegger in 1926 infamously described Man as a being unto death.[23] No ancient philosopher could have made such a statement. But, as Nietzsche had observed, the Europeans had finally killed off the Platonic God of Nature. After centuries of ambivalence toward Classical culture, Europeans simply disowned the super-sensory realm of Ideas, Forms, Nature, and Beauty. What was really troubling about this was that the "death of God" turned more or

---

20. See especially Kirk, *Roots of American Order*.
21. Parsons, *Structure of Social Action*.
22. Berger, *Scared Canopy*.
23. Heidegger, *Being and Time*.

less instantly into the "God of death."[24] This "new" God had long been latent in Europe's litany of negative theologies. But, in the totalitarian twentieth century, it now became an explicit force. The political theology of the Faustian God of negation, finally unchained from the constraint of Classical Nature, led Europeans into the pit of self-immolation. The God of Nihil was "the Nothing that horrifies man and displaces him from his usual dallying and evasions." This was a God that called humankind "to recoil in terror of annihilation and to be horrified by devastation." This was the Being that was Nothing—the Surplus that was Empty.[25]

From its European crucible, the doctrine of Nihil spread to Islamic and Japanese fascists in the 1920s, 30s, and 40s. But Americans, for the most part, were immune to it. Why this is so is an interesting question. Eager to explain away the catastrophe of Nazism, Heidegger pointed his finger at the God of metaphysics.[26] Platonism apparently was the cause of Europe's disgrace. Arguably holding a dead God responsible for European nihilism was intellectual dishonesty at its worst. If anything, the opposite was true. The God of Form, Nature, and Beauty has habitually inspired resistance in societies corrupted by nihilistic disorder.[27] America was able to resist the totalitarian plague because it was a *metaphysical republic*. In the settler societies of the New World, Classical Nature was never repudiated. Intellectuals toyed with the idea of doing so, but outside the universities this was never accepted, and even inside the universities voices like Heidegger's student Leo Strauss (who had fled Nazi Germany) could repeatedly assert that there is a Nature that is imperishable, that cannot be voided, nullified, or negated, and that cannot be put into question.[28] In America, the God of Nihil remained on trial. It had to answer to Nature's God.

---

24. Murphy and Roberts, *Dialectic of Romanticism*.

25. Heidegger, *Basic Concepts*, 42–65. *Basic Concepts* is a lecture series Heidegger gave in the winter semester of 1941. For a further discussion of this, see Murphy, "Pitch Black Night of Human Creation."

26. In lectures between 1936 and 1940, and again in 1943.

27. Eric Voegelin made the point that what counted about the metaphysics of Plato was not "Platonic philosophy" or "doctrine," but Plato's resistance to the disorder of the surrounding society and his effort to restore the order of Hellenic civilization through the love of wisdom. See Voegelin, *Plato and Aristotle*, 5.

28. See, for example, Strauss, "Jerusalem and Athens," 129; Strauss, "Three Waves of Modernity," 88.

Here we see the great opposition in modernity: between a nihilism that is self-devouring and a classical nature that has no beginning and no end. On one side of the divide is the necroromanticism of the Faustian Creator God. On the other side is the God glimpsed in the unlethal truth (*alētheia*) of the ancient Greeks—in Nature's implicit resistance to the necropolities of death. Even Locke and Hobbes, who reduced Nature to Natural Rights, agreed on this. Accordingly, a government that fails to protect its citizens or subjects from death, or that encourages movements of "thanatics," violates the *phusis* on which the state is erected. To define the human being as a being unto death, or society's work as producing the terror of annihilation, is to adopt the standpoint of the necropolis. Americans in the main found this repulsive. Instinctively, they asserted "life against death."[29] The best of their intellectuals warned of the appeal of the city of tombs. Lewis Mumford did so repeatedly.[30] What Americans sensed was the radical deterioration of the capacity of some societies to assert order in the face of chaos. The very point of such an order was that it did not spiral down through a lack of energy or spirit into depression, de-moralisation, and death-fixation. In American terms, this might best be described as an order of liberty. This is a kind of freedom, like the freedom of a great dancer or a great athlete, which is expressed through grace, balance, and limit.

NO GOING HOME

Americans are familiar with the difficulty of creating an order of liberty. Such an order is a paradoxical hyphenation of two seemingly contrary things: contingency and necessity. Liberty is the political expression of contingency. The American world is filled with contingency. This world—modern through and through—is the product of rights. Rights or permissions do not direct action but rather leave action to the choice of the individual. This is a social world where contingency and uncertainty are pervasive, and choice and election are inescapable suppositions for action and conduct.[31] American political ideologies venerate the "*I can*"

---

29. One of the rare European intellectuals to have taken up this viewpoint is Agnes Heller, who in multiple works insisted that "life" and "freedom" were the necessary axioms of any modernity that was not self-destructive.

30. See, e.g., Mumford, *City in History*.

31. One of the better known, not to say notorious, products of this Neo-Thomism was the work of the Toronto savant Marshall McLuhan. McLuhan was the most mercu-

and the "*we can*." But, in order that they "make sense," contingent acts—free acts—have to be integrated into a larger whole. Otherwise such acts end up being arbitrary or absurd. St. Augustine described this as the universe constantly sliding towards the abyss of nothingness. Cumulatively, acts that are arbitrary or absurd portend chaos. Chaos is the antonym of order, and it is order that signifies meaning. Human beings create meaning by organizing contingent elements and actions into patterns and systems.

For Americans, large-scale order is most visibly represented in the idea of a "new order of the ages"—the constitutional schema that the American Founders created. But underlying this is a still larger order of meaning. Religion is often used as a descriptor of this. Most Americans will tell you they believe in God. America is more conventionally religious and more church-going than any other modern nation and most pre-modern ones. But this sociological fact is less important than the metaphysical sense that permeates American society. This metaphysics is roughly equivalent to what the ancient Greeks called *phusis*. Emerson offered the most characteristic American reading of *phusis*. He called it the choral harmony of the whole. Notably the most perceptive European observer of the Americans, Tocqueville, missed the centrality of *phusis* thinking to America. The United States is everything that Tocqueville said it was. It is restless, rootless, vulgar, enterprising, improving, levelling, and teeming. But it also has an enormous capacity to turn turbulent energy into visible order without hierarchies or rules. Emerson identified the medium for extracting temperate order out of restless chaos: "Design. It is all design. It is all beauty. It is all astonishment."[32]

However such *phusis* is interpreted—whether it is understood as Nature, God, Beauty or Necessity—it is the intimation of something sacred. It is sacred not because it signifies a particular order of existence, but *because it signifies the very existence of order*. Alexander Pope put it beautifully: "Order is heaven's first law." Tom Paine spoke movingly of the "unerring order and universal harmony reigning throughout the whole" of nature—through self, society, and cosmos. What is sacred about this order is the equilibrium at the heart of it. Such equilibrium has a social

---

rial intellect of the Great Lakes ecumene, and is interesting for his ability to re-work what was at heart a Catholic worldview into an explanation and diagnosis of broadcast communication. He converted to Catholicism in 1937.

32. Emerson, "On the Relation of Man to the Globe."

pay-off. It is the force that orchestrates the flow of energy between social actors, social parts, and social systems. This flow stops or at least delays the onset of entropy. Balance thus secures negative entropy. When such order is torn asunder, we know that something terrible is at work. The opposite of order is chaos.

When societies, states, and empires slide into chaos, the question of civilization is posed. When this began to happen near the turn of the twentieth century—notably in the Balkans in the 1890s, and then in the First World War—Americans asked themselves: can we (should we) do anything to avert it? There were two possible answers to this question. One said that the descent into the abyss was the product of the vices of the Old World, and Americans should leave well alone. The other said that America had a responsibility to mend and restore, or replace, the sacred canopy when it had been torn asunder. One response was isolationist; the other was internationalist. But either response required justification to the world. By the beginning of the twentieth century America had become large enough and powerful enough that it had to answer for its actions—or its inactions. The problem was: in what "language" could it respond? Its home-grown ideologies were not much help to it. They were "domestic languages"—intelligible for local consumption. Foreigners found them opaque.

The thing that is striking about America's home-grown ideologies—the republic's broad-spectrum of liberalisms—is that they have had little resonance outside of the United States. Where variants of European socialism spread around the world, virtually no American natural rights ideology found followers abroad. The American way of talking about the experience of contingency translated badly. The idea of a Herbert Croly-style German or Egyptian "progressive" is virtually inconceivable.[33] While European natural rights philosophies proved to be highly exportable, not least of all to the United States, their incarnation in the various strands of American political ideology defied re-export. This posed a number of problems for a new world power: not just for the communication of its influence to others, but also for explaining itself, orientating itself in the world, and understanding the nature of others with whom it had to deal.

---

33. Croly, *Promise of American Life*. The kudos heaped upon this work is simply unintelligible to a non-American. See, for example, Rorty's encomium in *Achieving Our Country*, 46-49.

In 1955, the Harvard historian Louis Hartz (1919–1986) observed that liberalism had acquired a virtual stranglehold on the American mind. America's "irrational Lockeanism" had become "one of the most powerful absolutisms in the world."[34] Hartz noted that, as a result, America lacked a crucial combative horizon—something against which natural rights nostrums must struggle if America was to gain an adequate understanding of itself and of others in the world. Hartz's argument was that natural rights thinking had emerged in Europe in the modern age as an adversarial force in opposition to the thought of the *ancien regime*. In turn, natural rights doctrines had been strenuously challenged by feudal ("Tory"), socialist, and radical ideas. Hartz had the notion that a world-encompassing dialectic that might eventually provoke America into an intellectual engagement on a macro-historical scale. What caught Hartz's attention were idea fragments that had lodged in various places in the New World—the rural Calvinism of the Dutch (Boers) in South Africa, the reformist ex-English socialists in Australia, and Catholic Aristotelianism in French Quebec and Latin America. He never fully articulated it but he seemed to hope for a conversation, or perhaps an argument, of American natural rights liberalism with the feudalisms, socialisms, and radicalisms of the New World.[35]

At any rate, Hartz made a telling point: to answer the dilemmas of modern political life, America had to *step outside of its own self*. "Instead of recapturing our past, we have got to transcend it ... There is no going home again for America."[36] But if America could not "go home," then where and how could it find the intellectual *agon* that it seemed to lack? Hartz's reference to Catholic Aristotelianism appeared to give the most credible hint. This was not because it promised a re-run of the Catholic-Protestant divide that once had fired Europe. That *agon* was of another time and place. It had been convincingly extinguished in America by the natural rights doctrine that separated church and state, and guaranteed freedom of worship. Rather it was the case that Aristotelianism and Catholicism were as equally adept as Enlightenment natural rights ideas *in translating across time and space*. Like the Enlightenment, they had "no home." In order for America *not* to go home, its natural rights tradition had to be paired with some powerful contrarian currents.

34. Hartz, *Liberal Tradition In America*, 58; also 3–23, 58–60.
35. Hartz, *Founding of New Societies*.
36. Hartz, *Liberal Tradition in America*, 32.

The Chicago Encyclopaedists had similar intuitions. Certainly Robert Hutchins did. Hutchins' legacy was two-fold. In the 1930s, during his tenure as President of the University of Chicago, Hutchins promoted the works of Aristotle, Aquinas, and Cardinal Newman as paradigm instances of the Great Books,[37] and he encouraged around himself Neo-Aristotelians like Richard McKeon.[38] Later, in 1941, Hutchins co-founded the Committee on Social Thought also at the University of Chicago.[39] This founding was Hutchins's greatest legacy. No intellectual center in America ever attracted so many brilliant minds. Its early members included Leo Strauss (1949–1967), Saul Bellow (1962–1993), Hannah Arendt (1963–1967), and Friedrich Hayek (1950–1962).[40] For all of this, though, it was the Neo-Aristotelian experiment that was closer to Hutchins' own desire to create the bridge between Europe and America represented by the Great Books tradition—the epitome of Western Civilization.

Chicago Neo-Aristotelianism developed in parallel with Catholic Neo-Thomism.[41] Both understood that form was the necessary correlate

---

37. Hutchins, *Higher Learning in America*. See his comments on Aristotle (pp. 56, 68, 81, 84, 97–98, 103, 119), Aquinas (pp. 63, 96), Newman (pp. 63, 103), and Plato (pp. 78, 81, 84).

38. McKeon, *Selected Writings of Richard McKeon*. Or as the quip went, at the University of Chicago Jewish professors taught Catholic thought to Protestant students.

39. He did this with the historian John U. Nef, the economist Frank Knight, and the anthropologist Robert Redfield.

40. This concentration of mercurial talent is unlikely to be repeated. Even distinguished later Committee alumni like Allan Bloom, Paul Ricoeur, Charles Rosen, and Leszek Kolakowski do not quite measure up to Chicago's great "Attic" moment. Mainstays of the Committee through its great period included the classicist and translator David Grene and the sociologist Edward Shils.

41. When migrants from Eastern and Southern Europe first arrived in the United States in the 1870s and 1880s, the dominant Catholic theology had been a sentimental gothic-romanticism that looked back to an idealized medieval world. However, an anti-romantic Thomism emerged in reaction to this at the turn of the twentieth century. Etienne Gilson—at Toronto—and Jacques Maritain were representative figures. Leo XIII's 1879 encyclical *Aeterni Patris*, calling on the Church to study St. Thomas, triggered the movement. Fordham University (in New York City) and St. Louis University were important Neo-Thomist centers. *Modern Schoolman* (1925), *Thought* (1926), and *New Scholasticism* (1927) were its chief journals. (Gleason, *Keeping the Faith*, 14–29, 113–14, 140–42, 148, 167.) The great modernist architect, Mies van der Rohe, was one of the fellow travelers of this Neo-Thomism. Based in Chicago, Mies developed one of the most influential international architectural practices in the 1940s, 50s, and 60s. Murphy & Roberts, *Dialectic of Romanticism*, 137–44.

of freedom, and that contingency required an encompassing order. Yet their understanding of this was curiously abstract. Both confused history with the history of reading texts. The orthodox Neo-Thomist account of Being is inseparable from "acts of interpretation" of the Great Books of the Classical-Christian corpus.[42] This is ironic given the traditional Catholic bias towards visual and audile culture. Neo-Thomism ended up mimicking the Hebraic-Protestant passion for texts.[43]

LIQUID GEOGRAPHY AND SACRED CANOPY

Neo-Thomism and Neo-Aristotelianism were too bookish, and the counter-Lockean currents of the New World were too distant or too antediluvian, to be more than curiosities to Americans.[44] One interesting thing of lasting value that Hartz's thesis about the New World did introduce, though, was the figure of geography. Hartz concluded that location was a determinant of political ideology. It mattered whether

---

42. Unorthodox Neo-Thomisms were conceivable, though. A left-field case was that of Marshall McLuhan (1911–1980). A Catholic convert, McLuhan spent a teaching career at the Jesuit and Thomist-dominated St. Louis University and at St. Michael's College at the University of Toronto. McLuhan blended an eccentric Catholicism with the brilliant theories about media dreamed up by his older, Baptist-raised, Toronto colleague Harold Innis (1894–1952). Gilson, a colleague of McLuhan's at Toronto, could not make sense of McLuhan's tangential Catholicism. Innis' chief insight was that a wide and interesting range of media—from the plastic media of architecture to the network media of roads and railways—were as decisive in human history as texts. The medium was the message, as McLuhan later glossed. Though both were raised on the Canadian plains, Innis was an intellectual product of Chicago in the same way that McLuhan was of St. Louis. Both were in those places for relatively short times—Innis to do his PhD in Economics at the University of Chicago, McLuhan as a junior professor in St. Louis (1937–1946). Innis received his PhD degree when Robert Park and George Herbert Mead were both teaching at Chicago.

43. To compound the irony, American Protestant evangelicals rejected texts and embraced dramaturgy, rhetoric and music.

44. The twentieth-century iteration of Thomism lost its impetus in the 1960s. Romantic-liberationist currents pushed it aside. An ambitious version of Catholic Aristotelianism was to appear later in United States in the 1980s—with the publication of works by the relocated Scottish Marxist, Alasdair MacIntyre. These works self-consciously agonized the Enlightenment tradition by setting against it the rich history of Aristotelian Christian thought. This attempt to agonize Enlightenment liberalism had its limits. For one thing, it was a very bookish reading of history. It replicated the bias of the Chicago Encyclopaedists that "the West" or "Civilization" was a chain of texts—each one supplementing texts that had come before. The equation of text and "meaning in history" was problematic. Painting, sculpture, architecture, urbanism, design, and performance had a curiously peripheral role in this definition of civilization.

political ideologies came from Quebec or from New England, from the American Mid-West or from the American West. More important still than the specifics of Hartz's work was his general approach. The reader of his classic work *The Founding of New Societies* is invited to look at history and politics through the lenses of space and geography. This is valuable because it helps us begin to answer the question of civilization ("what is it?") in ways that the Great Books approach cannot do and does not do.

Space in the guise of geography is, and always has been, a major determinant of civilization. It conditions what civilization is and where it emerges. The question of civilization at its heart is a question about order and chaos. Certain geographies—that is also to say certain arrangements of nature—tacitly cooperate in the creation of order. They become Nature's—or the Cosmos's—correlate of social order: an aid to the creation of social meaning and an impediment to its dissipation through entropy. Sea regions, littoral topographies, and the liquid geographies of rivers and coasts are particularly important in the creation of civilization. Liquid geographies are nature's "breath" (*pneuma*). In the course of human history, these have been amongst the most important spaces of passage. They have been the spaces that have regulated the most intense comings and goings, giving and receiving, entries and exits of human beings.[45]

This pneumatic circularity lies at the core of Being. The rhythm of being is inscribed in the nature of civilization. Social systems dissipate entropy or disorder, at least for a time, through the interactions (giving and receiving) between system and environment.[46] "Breath" or *pneuma* is a model or metaphor for an alternating, dyadic rhythm. This is nature in the sense of the Greek *phusis*. In so many of its aspects, civilization is an artifice. It is the work of human ingenuity and design. But this human ingenuity is always conditioned by nature. Nature aids design. "Obey nature," advised the Stoics. In practice, autopoietic civilization does follow nature.

One of the indubitable aspects of nature is space. From the standpoint of civilization, the most interesting kinds of spaces are those of passage and traffic, circulation and revolution. Such spaces exist only in the abstract—as pure potential—until human beings make something of

---

45. Murphy, "Marine Reason," 11–38.
46. Clarke, *Person, Being and Ecology*, 48–49 and Ibana's commentary, 91.

them: until, for example, the American pioneers crossed the Appalachian Mountains to settle in the Ohio and Mississippi Valleys—blessed with the great river systems that connect the Great Lakes system with the Caribbean Sea. In doing so, these settlers, unconsciously, were enacting a "sacred" space: one that had its own kind of implicit rhythmic order. This was a space where human beings acted in a mimesis of nature. It was the Sphinx of the American Revolution, Thomas Jefferson, who understood the macro-temporal and macro-spatial significance of this. After all, he purchased precisely this "geography of connection" from the French. He saw it as a central to the creation of an "empire of liberty" that would evade the chaos of European history: the chaos caused by Faustian despotism.

Jefferson's intuition was supported by precedent. For even across the Atlantic, the great examples of civilization were creations and creatures of liquid geographies: the Mediterranean-Black Sea region and the North Sea-Baltic Sea region—and the fingers of the rivers that drained into them. From these littoral regions had arisen classical antiquity, the Renaissance, and the modern civic capitalism of North-West Europe. What had existed on the maritime periphery of Europe was multiplied in the case of North America. For Nature had blessed North America with a historically unprecedented series of sea and littoral regions suited to intensive commercial, civic, and intellectual transactions: the Eastern Seaboard (Boston, Philadelphia, Washington), the Hudson–Great Lakes' region (New York City, Detroit, Toronto, Chicago), the Mississippi–Gulf–Floridian Peninsula region (St. Louis, New Orleans, Houston, Miami), the California Coast region (San Francisco, Los Angeles), and the Puget Sound region (Seattle, Portland, Vancouver).

Those who like irony should relish the fact that world history is regional. It is the creation of littoral city regions with world reach—in exactly the sense that Venetian traders and bankers impacted economies and societies from the Baltic to the Silk Route. That America, by the beginning of the twentieth century, had become world historical is without question. What is more intriguing to observe is the foundation upon which America's ascent to world history was achieved. It is interesting to compare America with Athens and Rome. These states launched themselves from *one* sea region (the Mediterranean) or in the case of the classical Greeks it is perhaps more accurate to say *two* sea regions: the Mediterranean and Black Seas—fighting over the command of the passage between those two seas provided the basis for Greek cycle of epic and tragedy rooted in the story of the Trojan war. The rise of American power and civilization was based not on one or even two but on at least *five* sea regions. (*Six* if we add the historic case of Canada's Hudson Bay.) One of the things that make Thomas Jefferson great is that he had inklings of this. He intuited that America eventually would cross east and west and north and south from coast to coast.[47]

Like many of his generation, Jefferson was schooled in classical history and philosophy.[48] It is difficult for casual observers to appreciate the modernity of this. What gave Jefferson such an interesting view of macro-history is precisely that this was history not understood from the standpoint of territorial Europe but from the standpoint of the littoral. American models of government and civilization came from the maritime periphery: from the ancient and Renaissance Mediterranean, and from the North Sea powers, the British and Dutch, and their notions of Commonwealth and Republic.[49] Antiquity was especially productive for thinking about social models independent of the realpolitick of the day.

We find the sympathy that Jefferson and his Deist peers had for Classical archetypes resurrected in the twentieth century, especially amongst the brilliant cohort of thinkers concentrated around the Committee on Social Thought at the University of Chicago. This cohort did what Jefferson had done: they forged an astonishing and paradoxical worldview of modernizing classicism. As in the case of Jefferson, this allowed them to take on board and yet at the same time qualify the instinctive natural rights ethos of America—to make of nature something more than inalienable rights. To do this required a philosophical history—or rather a philosophical geography—of America that was not dependent on the history and geography of Europe in Charlemagne's sense.

---

47. In a remark to James Monroe in 1801, Jefferson expounded his view: "However our present interests may restrain us within our limits, it is impossible not to look forward to distant times, when our rapid multiplication will expand it beyond those limits, & cover the whole northern if not the southern continent, with people speaking the same language, governed in similar forms, and by similar laws." See Alstyne, *Rising American Empire*, 87.

48. Murphy, *Civic Justice*, 284–89.

49. The geopolitical dominance of these states was based on naval power. The both lacked the tradition of large standing armies that could overwhelm state and society.

To play the classicist as Jefferson did might be thought a curiosity, yet it implied a skeptical nod toward a history and geography (the history and geography of Old Europe) that many, perhaps most, Americans saw themselves as being an exception to. Even Americans who were vociferously pro-Western in international policy had doubts about the catastrophes of the European past. The West seemed to them to be more like a burden to be carried than a model to be admired. The views of the latter-day Chicagoans, contemporaries of Louis Hartz, agreed with Hartz at least in this—that there could be "no going home again for America." As we have seen, Hartz fished around the New World for agonistic partners for American liberalism, a strategy not without merit. Hartz intuitively identified civilization with settler societies. This was a brilliant intuition, but one that he never fully developed.

The latter-day Chicagoans looked elsewhere—to the historic, multilayered civilizational seedbed of the Eastern Mediterranean.[50] There is an argument to be made that, already in antiquity, the Eastern Mediterranean was one of the principal historic progenitors of settler societies.[51] It laid down a template for all later settler states. In that sense, there was a tacit convergence of the latter-day Chicagoans and the Hartz thesis. But this convergence is only to be seen in hindsight. The Chicagoans' explicit preoccupation was the Eastern Mediterranean—its philosophical, political, and religious history. This was a mirror against which North Americans could measure themselves and their role in the world. What the 1950s and 60s generation of Chicagoans proposed, in effect, was a dialogue between the shores of the Great Lakes and the ecumene of the Eastern Mediterranean. William McNeil furnished a series of world historical studies of the terrain between the Crimea and Venice—spanning the Venetian and Ottoman Empires and Modern Greece.[52] Leo Strauss defended classical nature, and questioned modern natural right. With a Socratic gesture, he ironized the American idea of the liberal—turning him into a Greek gentleman. Hannah Arendt sketched America's debt to Greece and Rome,[53] and offered a non-Lockean account of the human

---

50. The term "seedbed," used of the Eastern Mediterranean, is Talcott Parsons's. See Parsons, *Societies*, 59

51. Murphy, *Civic Justice*, 19–20.

52. McNeil, *Europe's Steppe Frontier*; McNeil, *Venice*; McNeil, *Metamorphosis of Greece since World War II*.

53. Arendt, *On Revolution*.

condition.[54] In *On Revolution*, she presented the first great philosophical account of the difference between Europe and America. Friedrich Hayek famously defended Anglo-American self-organization against European centralism. English Old Whig ideas stood in for the Greeks in Hayek's social philosophy.

OPEN SYSTEMS

Chicago was a prism for a powerful idea. This was the notion that American self-understanding was best served by looking in the civilizational mirror of the Eastern Mediterranean. Hannah Arendt exemplifies the prismatic quality of Chicago. She was the paradigmatic New York intellectual but her consummate work, *The Human Condition*, came out of the Walgreen Lectures she gave at the University of Chicago in 1956. The case of Eric Voegelin echoes this. Working away in that faded residue of former littoral power—Baton Rouge in Louisiana—he developed reflections "on the form of the American mind"[55] into a long meditation on the "order and history"[56] represented by the Mediterranean ecumene. The first mature statement of Voegelin's philosophy, *The New Science of Politics*, was created during a short tenure at the University of Chicago, where he delivered the 1951 Walgreen Lectures.[57] Similarly, in Toronto in the early 1950s, Harold Innis' mercurial studies of modern communications and empire drew heavily on comparisons with the Greco-Roman

---

54. Arendt, *Human Condition*.

55. Voegelin was born in the Rhine ecumene, in Colonge in 1901, and spent three years on a Rockefeller scholarship in the United States (1924–1926), after which he wrote *On the Form of the American Mind*. In 1938 he returned to the U.S., in flight from Nazism. He was professor at Louisiana State University (1942–1958), moving to Munich in 1958, and returning to the United States in 1969 to Stanford University. His *Collected Works* were published in 1989.

56. Voegelin, *Order and History*. Vol. 1. Israel and Revelation. Vol. 2. The World of the Polis. Vol. 3. Plato and Aristotle. Vol. 4. The Ecumenic Age. Vol. 5. In Search of Order.

57. Voegelin, *New Science of Politics*. The Walgreen Lectures, along with the Gifford Lectures, were probably the greatest lecture series in history. Yves R. Simon presented them in 1948, Leo Strauss in 1949, Voegelin in 1951, Robert Dahl in 1953, and Arendt in 1956. They were the basis for Simon's *Philosophy of Democratic Government*, Strauss' *Natural Right and History*, Dahl's *Preface to Democratic Theory*, Kennan's *American Democracy*, Potter's *People of Plenty*, Maritain's *Man and the State* and Arendt's *Human Condition*.

past.[58] The prism of Chicago again was notable. Innis had completed his PhD at the University of Chicago.

Chicago was not omnipresent. Lewis Mumford regarded the Greek *polis* as the measure of civilization but owed this view to a New York education and to the early influence of Emerson, not to mid-Western Hellenism.[59] Yet Saul Bellow was still right to note the peculiar power of Chicago over the life of the mind. The case of the Austrian expatriate Ludwig von Bertalanffy illustrates this perfectly. In 1937–1938, Bertalanffy gave his first lectures on General System Theory—at the University of Chicago.[60] He was a little different, though, from most of the post-war Chicagoans. Bertalanffy was deeply influenced by the macro-historical thought of the 1920s and 1930s: Toynbee and Spengler. He wrote little on the triangulation of Athens, Rome, and Jerusalem. Yet his work—or rather his central thesis—is crucial to understanding why the Eastern Mediterranean rather than the West made good sense as a mirror for American self-understanding of the nature of civilization.

Bertalanffy introduced the notion of the open system.[61] He coined the term for biology, and then applied it across the spectrum of natural and human sciences. Biological organisms, he noted, were systems of elements. Bertalanffy distinguished two kinds of systems. The first kind, the (traditionally conceived) physical system, does not exchange matter with its environment. The system is closed. Without the import and export of matter, the system gradually runs down or breaks down. Closed systems invariably suffer entropy. The fate of a closed system is disintegration and death. Organisms in contrast can evade entropy—at least up to a point—by importing and exporting matter. This is the definition of an open system: matter flows across the boundaries of the system.

Bertalanffy applied his distinction between open and closed systems to many kinds of systems—including social systems. Indeed, in the case of social systems, the application is very apt. Most societies and states suffer from entropy at some time or other. Some suffer this more than

---

58. Innis' friendship with his Toronto classics colleague Charles Cochrane was an important influence in this respect. See Cochrane, *Christianity and Classical Culture*.

59. Sir Alfred Zimmern's 1911 study *The Greek Commonwealth* was also an early influence. Zimmern (1879–1957) was a classicist who became the first professor of international relations at the University of Wales.

60. Like Voegelin before him, he was a Rockefeller Fellow.

61. Bertalanffy, *General System Theory*.

others. In Bertalanffy's own lifetime (1901–1972), this was spectacularly true of the Eurasian empires: the Austro-Hungarian, German, Russian, Ottoman, and Soviet Empires. Each ended in breakdown, disintegration, and dissolution. Sometimes the run-down of a social system expresses itself in crisis and the incapacity for reform. Other times entropy-death, manifest in the commission of unnatural crimes, becomes its very rationale for existence. In the latter case, the bleakness of traditional social suffering is exceeded by a new, infernal kind of suffering that beggars the imagination. We see the latter exemplified by the fratricide of Europe in the First World War, the Nazi death camps, the Soviet Gulag (20 million dead), Idi Amin's Ugandan dictatorship (1971–1979, 300,000 dead), Pol Pot's rule in Cambodia (1975–1979, 1.7 million murdered), Radovan Karadzic's and Slobodan Milošević's genocide in ex-Yugoslavia (1991–1995, 430,000 killed, 3 million displaced), the fratricidal Iraq-Iran War (1980–1988, 1 million killed) and the 35 years of Iraq's Baathist necropolis (another quarter-to-a-half million dead and 4 million exiled), or Africa's civil wars at the millennium's end: the Democratic Republic of Congo four decades on from 1960 and 2 million dead,[62] a grisly bookending of King Léopold II of Belgium's late nineteenth-century heart of darkness where at least 5 million Congolese were killed.[63]

One of the reasons that America looked upon itself as an exceptional society was its sense of repulsion at the history of social entropy in Europe and elsewhere. In 1904, the African-American leader Booker T. Washington observed of King Léopold's deeds: "There was never anything in American slavery that could be compared to the barbarous

62. To cite just one of many other cases: Between 1989 and 1999, the pairing of General Omar al-Bashir and the Sorbonne-educated theological hard-liner Hassan al-Turabi were responsible for the deaths of some one-and-a-half million of Sudan's animist-Christian black African population. In 1994–1996, Osama bin Laden had his base of operations in the Sudan. Nothing changes. The British fought the earliest iterations of this war in Egypt and the Sudan in the 1880s and 1890s—protecting Coptic, African, and Greek Christians and others from Mahdi-inspired slaughter and enslavement.

63. Those—like Hobsbawm, *Age of Extremes*—who thought the twentieth century to be defined by its "shortness" were wrong. The age of extremes began well before 1914 and has continued unabated since 1991 and the collapse of the Soviet Union. While it may have been psychologically tempting to draw a curtain on the horrors of the twentieth century once the disgraceful Soviet regime expired, the same horrors were promptly repeated on Europe's door-step, in the Balkans, through the 1990s. The European response to Balkan genocide was to wring hands, negotiate with a recalcitrant power, or do nothing. It was the Americans who took military action and ended the plunge into barbarism.

conditions existing today in the Congo Free State." Robert Park—later to become the great Chicago School sociologist—campaigned with Washington against the Congolese killing field. In their public agitation through the Congo Reform Association, the pair paid homage to the opening paragraph of the American Declaration of Independence and its assertion of the inalienable right to life. Against the background of repeated crises of civilization in Europe and European-controlled societies, this was an affirmation of the universal law against murder and the slaughter of innocents. It was also an affirmation of the American metaphysics of order.

Civilization is a synonym for puzzling about social entropy. American interest in this question picked up around the First World War. This was partly because of the great European fratricide—and the question of whether America should intervene to stop it. Partly, also, American interest was raised because of the emergence in the late nineteenth century of movements that had begun to employ terror not only as the means but also increasingly as the end of political action: these movements appear in different guises, again and again, from turn-of-the-century Russian Slavophiles and Balkan irredentist nationalist terrorists through the Stalinist wing of the Bolshevik Party to the Italian Red Brigades and the German Baider-Meinhoff Gang in the 1970s to Islamist terrorists in the 1990s and 2000s with their mix of fascist ideology, anti-Semitism, and lurid pan-nationalist fantasies.

Thanatocracy—rule by death—poses dilemmas. There is the temptation of normal states to back thanatocracies because of the imperatives of real politik. Take one example: Anastasio Somoza Debayle's "anti-communist" regime in Nicaragua (1959–1980)—it killed possibly 50,000 of its own citizens. It took the United States government till 1978 to suspend its military aid to the regime. In the 1960s and 1970s, the low point of U.S. diplomacy, the American state regularly backed murderous dictators—from Suharto in Indonesia through Pinochet in Chile to Marcos in the Philippines. The long-term damage caused to these societies by tyranny was enormous. Realists, like U.S. Secretary of State Henry Kissinger, invoked European theories of international relations that relied on despots to contain a greater despotism (Communism). Romantic critics of this sordid realism replied with the adoration of Communist despotism. The lesson of the twentieth century was that such politics only encourages thanatocracy to do its worst.

Thanatocratic classes, though, will wage vicious wars to stay in power. So withholding support from or overthrowing a thanatocracy requires acute political and military judgment. Such political and military judgment rests on a more fundamental judgment: at what point can we determine that a state has fallen into the bestiarium?[64] Even more difficult is to understand how such regimes can be prevented in the first place. How can we stop a society sliding into the black hole of entropy and death?

Bertalanffy gave an interesting answer to this question. He had grown up with Spengler and Toynbee as early influences. He never lost his affection for their works.[65] Both thinkers were products of the twentieth-century crisis of European civilization. Spengler foresaw the end of the Faustian West. The first six volumes of Toynbee's nine-volume *A Study of History* were published between 1934 and 1939—as Europe plunged into barbarism. Bertalanffy's answer to the question of how social entropy could be avoided was more theoretical rather historical but it was no less cogent for that: societies avoid entropy by creating themselves as open systems. Societies that import and export matter across boundaries can resist entropy. The flipside of resisting entropy is the capacity for systemic organization at ever-higher levels. Open systems typically move in the direction of greater complexity: the greater the negative entropy of a social system, the greater the number of parts of the social system that can be integrated with each other. Negative entropy, as Bertalanffy observes, can be considered as a measure of order and organization.[66] Order is the opposite of chaos. Social systems in decline are subject to chaos. Breakdown, disintegration, and dissolution take effect through chaos.

According to Bertalanffy, order is generated when import-export occurs between a social system and its environment. This helps us understand why the latter-day Chicagoans were right to single out the Eastern Mediterranean as a key to American self-understanding of civilization. It was not because Athens and Rome and Jerusalem were the home of many Great Books—even if they were. It was rather that this liquid region, at its greatest, functioned with very high levels of transaction

---

64. On the political theory of the bestiarium, see Feher and Heller, *Eastern left, Western left*.

65. Bertalanffy, *Perspectives on General System Theory*, 74–84.

66. Ibid., 111.

between social system and environment. To put this in more concrete terms: the Eastern Mediterranean, for much of its history, was a portal space or ecumene. It was dominated by intensive exchanges between its great portal cities and littoral city-regions: Piraeus, Ostia, Venice, Genoa, Pisa, Constantinople, Alexandria, Rhodes, Antioch, and many others.

Social systems that are open continually traffic with their environment: what could be a better description of a portal? A portal or an ecumene is the space where import-export occurs. This space, or this space-time, necessarily corresponds to an open system. An open system has weak borders but strong order. How can we understand this apparent paradox? There are two ways that a social system creates borders: through rules and hierarchies, and through self-organizing order. Rules and hierarchies function to create order but they do not institute self-organizing forms of order. Rules and hierarchies require permissions and authorities to function, and the systems that they organize suffer entropy eventually. Self-organizing order in contrast is abstract and intuitive. It relies not on permissions and authorities but on mathematical-geometric principles such as balance, homology, oscillation, proportion, and symmetry. Bertalanffy thought that one of Spengler's great insights was that social orders have a mathematical foundation.[67] "Geometric" principles are crucial to the processes that self-organize the interrelation of the parts of a system or the relations between system and environment. The genius of such principles, which we see exemplified in the structure of the American Constitution, is that they function quasi-independently of social actors having to make explicit decisions or give explicit commands. Rather than organizing a social system exclusively on the basis of directions and rules, statuses and rights, in the case of open systems self-organization—operating via pictorial-aesthetic-mathematical models and schemas—plays a crucial role in securing the inter-relationship of social parts.

Bertalanffy supposed that in a closed system something like rules or hierarchies ("laws of nature," "evolutionary selection") could create order but that open systems were more dynamic because they were self-organizing. The city-states of the Eastern Mediterranean were the first to develop theories of how self-organization works through proportionality, symmetry, rhythm and similar "geometric" qualities. From these city states came a deep understanding of the civilizing forces of grace,

---

67. Ibid., 82.

balance and equilibrium. Taking his cue from Aristotle, Bertalanffy spoke of holistic systems where the whole is greater than the sum of the parts. The organizing force of the whole is neither additive nor is it a function of rules or codes. Rather, qualities like parallelism, analogy, allometry,[68] lattice and other kinds of branching and clustering, crystalline and meandering patterns provide the organizing force for macroscopic holistic interactions.[69]

ECUMENE AND THE MIRROR CITY

The nub of the argument is this: to understand the nature of portals, we need to understand the nature of organization without rules or social hierarchies. We can give various names to such organizing forces, not least of all autopoiesis or self-organization. We can also find many examples of this. When we are dealing with social self-organization, some of the most powerful examples occur in portal societies. Reflexive self-analysis of the nature of such societies began in the Classical Mediterranean.

In simple terms, the Eastern Mediterranean ecumene is a model of the water-bound ecumene in general. Even in the modern era of oceanic power, sea regions still remain crucial entities. We cannot imagine the idea of modern natural rights, or modern civic capitalism, without the contribution of the English-Scottish-Dutch North Sea ecumene. In the same way, the economic modernism of twentieth-century Hong Kong, Singapore, Taiwan, and Shanghai depended on the transactions of the China Seas. We cannot understand the birth of American intellectual life, unless we take into account that it emerged in the littoral world of the Eastern Seaboard and the North American Great Lakes, with their migrant cities and conjugations of nations and nationalities, and their intense rhythms of import-export. This is symbolized neatly by the pragmatism of John Dewey and George Herbert Mead—products of the Great Lakes hugging University of Michigan and University of Chicago, the antipodes of the littoral worlds of Baltimore and Boston from which the work of William James and Charles Sanders Peirce emerged.[70]

68. Denotes the growth of a part of a body at a different rate from that of the body as a whole.

69. See, for example, Bertalanffy, "Evolution: Chance or Law" in *Perspectives on General System Theory*, 137–48.

70. It should not surprise us either that the longest term of employment that the Peirce had in his difficult life was with the U.S. Coast and Geodetic Survey.

The ecumene is a liquid region. It is a liquid space of intersection between social system and environment. Within that region, a particular type of economic, political, and cultural interaction is possible. This can be best grasped with the concept of the "mirror city."[71] This idea points to the propensity of major cities in a sea ecumene to become "mirrors" of each other. The history of interaction between portal cities produces porous relationships in language, culture, commerce, and residency. The Mediterranean in both the ancient and modern eras produced well-known examples of mirror cities—for example, Athens and Alexandria in the Hellenistic era—or ones like Marseilles and Algiers today that may fall below the threshold of common awareness but nevertheless remain influences on long-term geo-social, geo-political, and geo-intellectual dynamics.[72] The long-lasting and deep bond between Constantinople and Venice in the Byzantine and Ottoman eras is a spectacular example of a mirror relationship. To date, interactions across oceanic spaces, even with the relative speed of twentieth-century communications and transportation, remain too diffuse to create oceanic mirror cities. But, in more compact sea (or sea-like) spaces, relationships of mirroring continue to be created—with powerful effects.

Chicago is a classic example. Historically, Chicago had three mirrors: Toronto, which faces Chicago from across the waters of the Great Lakes, Detroit which looks northward across the liquid border to Canada, and New York City.[73] Their history of linkage is complex, not

71. Blank, "The Pacific," 265–77.

72. Blank puts it thus: "The flood of European immigrants into Algiers during the colonial period was reflected by a reverse flow of Algerians into Marseilles in the decades after World War II. North African enclaves in the European city mirror European enclaves in the North African city. In some neighborhoods of Marseilles, it is difficult to tell whether one is Europe or North Africa. Before independence, the same could be said of Algiers" (Blank, "The Pacific"). On the mirror relationship between Marseilles and Algiers, see Murphy, "France's Mediterranean Antipodes".

73. The mirror of Chicago and Toronto is very apparent in the life of mind. As a counter-point to Dewey, Arendt, Strauss, Hayek, and McNeill at Chicago, we have Harold Innis, Eric Havelock, Marshall McLuhan, and Jane Jacobs at Toronto. For the work of the Toronto intellectuals, see: Innis, *Empire and Communications*; Innis, *The Bias of Communication*; McLuhan, *Gutenberg Galaxy*; Havelock, *Liberal Temper in Greek Politics*; Havelock, *Preface to Plato*; Havelock, *Literate Revolution in Greece and its Cultural Consequences*; Jacobs, *Economy of Cities*; Jacobs, *Cities and the Wealth of Nations*. On Innis and McLuhan's life and work, see Heyer, *Communications and History*, chapters 8 and 9. For Innis, see Acland and Buxton, *Harold Innis in the New Century*; Havelock, *Harold A. Innis*; Connor, "Harold Innis." On McLuhan, see Terrence, *Marshall McLuhan*.

least in the case of New York. New Yorkers were instrumental in financing the development of Chicago in the nineteenth century. Chicago in turn was instrumental in getting the Erie Canal built—allowing a continuous fluid passage between New York City on the Hudson River and Chicago on the Great Lakes. When the Canal was completed in 1825, it turned New York City's harbor into America's number one port. The Chicago-New York mirror, with its antipodes and affinities, was echoed later on in the twentieth century—by the relationship between the San Francisco-Oakland-Bay Area and the mirror city-littoral region of the Southern Californian strip-polis that stretches from Santa Barbara to San Diego-Tijuana. Today, as wealth and population moves towards the Gulf of Mexico-Floridian Peninsula region, Houston and Miami increasingly play the role of mirror cities.

The principle assumed here is that thalassic regions are powerful sources of the "rule-less" order of grace that resists the entropy-death of societies. Of course there are examples of thalassic failures. New Orleans is a case in point. The ineffectualness of its city officials, the lack of self help, and the urban lawlessness graphically revealed during Hurricane Katrina in 2005 was the result of long-term social entropy. In general, however, American portals generate more energy than enervation. The dynamics that lifts drive and activity above depression and violence can be described in the following terms: thalassian circumstances create a demand for seaports (and airports and rail ports). Portal cities become nodal points for the contact, transmission, exchange, transformation, and finally supersession of rules and roles, hierarchies and statuses, cultures and worldviews. These are places known especially for their diaspora, exiles, traders, pilgrims, explorers, migrants, "circulating" administrators, "journeymen," and traveling artists. Portals encourage communication across cultures and customs. The liquid space of the ecumene is a medium for replacing norms and rules, social hierarchies and chains of command. If this was the case historically in classical antiquity, then the modern age of civic capitalism, if anything, intensified this condition. As sea regions came to be incorporated into the global system of oceanic power—and, correspondingly, as the United States developed on the unprecedented foundation of multiple sea (and sea-like) regions—this had the effect of compounding the concentration of habits, cultures, stories and worldviews in the liquid spaces of portal cities.[74]

74. Murphy, "Ethics of Distance"; Marginson, Murphy, Peters, *Global Creation*, chapters 2-4.

Park, Mumford, Arendt, Voegelin, and Strauss all treated the city as a suggestive model for American power and civilization. Their conceptualizations of this drew on a systemic analogy with the classic maritime polis. This was a good analogy. Ancient Athens was a point of interaction for Greeks from Southern Italy, Attica, and Asia Minor. Later, Alexandria brought Jews and Greeks together, Rome synthesized Latin and Greek cultures, and Antioch melded Jews and Gentiles into Christians. This was enabled by open systems of port cities and liquid regions. What made this possible was not simply the pasting of bits and pieces of culture and habits together. That is not civilization. Civilization assumes coherent meaning; coherent meaning assumes systemic pattern; systemic pattern assumes an ordering principle. That is really what open systems or portals do: they are the medium through which order is generated. Order is generated when societies and cultures meet—when Greek city meets Greek city, Dorian meets Ionian, Greek meets Roman, Hellenized Jew meets Orthodox Jew, Jew meets Gentile, Venetian meets Byzantine, Ottoman Greek meets Ottoman Turk—under certain conditions. If we live in a place where there is traffic between different normative cultures and where there is an ecumenical "reason of grace" that becomes a container all of them, then a common order is possible. The city—from Athens to Venice to Chicago—is a way of thinking about how such a thing is possible. The great portal city is both the symbol and the embodiment of an ecumenical order.

How does an ecumene create order? It does so by triggering negative entropy. To do this requires self-organizing capacity. Another term for this is pneumatic spirit. This simply means that ecumenical order is created not by instituting new hierarchies or new rules, new norms or new statuses, but by other kinds of media. Without question, virtue ethics and divine commandments, religious laws and secular legislation, liberal rights and libertarian permissions are media from which societies emerge. But they are not the only kinds of media. There are also nonlinguistic visual, tactile, and kinetic media, mathematical-geometric and design media. These are form-creating media. Norms and rules, hierarchies and statuses employed on their own create closed systems. Designing media—media that are architectonic—create open systems.[75]

We see this luminously represented in great portal cities at their height. They habitually exhibit a plastic genius for the organization of

75. Murphy, "Architectonics"; Murphy, "Communication and Self-Organization."

matter. They are brilliant builders of churches, mosques, universities, hotels, stock exchanges, stations, offices, and the rest. Such building is the outward expression of spirit (*pneuma*). Spirit is the great antagonist of entropy. Spirit sweeps entropy aside. Spirit is negative entropy. Entropic social systems lose their capacity to organize matter. Spirit is exactly this capacity. One of the basic reasons for the loss of spiritual capacity is that norms and rules, statuses and hierarchies—no matter how sophisticated—on their own create non-porous boundaries between a social system and its environment.

The consequences of this are two-fold. Matter flowing between a social system and its outside is reduced. The obverse of this is that rules and hierarchies create inward-looking and de-spirited order. The over-all effect is a dampening of human activity.[76] Depression, de-moralization, and disintegration follow. To overcome this, an order that ensures porous boundaries is necessary. This order bridges between system and environment. It does this not by rules and hierarchies but by equilibrium, balance, symmetry, and other architectonic forms. It organizes system and environment relations through patterns that are self-organizing. Such patterns can be extended in unpredictable or spontaneous ways, and in ways that do not necessarily require explicit authorization. This kind of order is pneumatic.[77] Another word for it is sacred order—self-organizing order that invests human life, and societies in history, with meaning and coherence.

76. Or as Innis put it in his notebooks (1947–1948): "Closed systems result of written tradition, i.e., Spengler-Toynbee. Totalitarian states—belief in power of written word of government" (*Idea File of Harold Adams Innis*, 123).

77. For further discussion of the nature of pneumatic order, see Carroll, *The Western Dreaming* and Murphy, "Marine Reason."

9

Perceiving *Freedom* and *Meaning* in Nature

Operationalizing Trans-Classical Systems Theory for Converging Scientific and Religious Knowing[1]

MARKUS EKKEHARD LOCKER

ABSTRACT

The present chapter considers the question of whether nature's processes pursue a given goal or emerge through self-generated occurrences, by pointing to the fact that both science and religion view nature and all creatures—beside the human person—as irrational objects. Maintaining this epistemological caesura between the knowing subject and the known object will indeed not allow for any other conclusion than that intelligible developments in nature are either caused by God, or the result of accidental events.

In the search for a third view that is acceptable to both science and theology this essay argues that Trans-Classical Systems Theory [TCST], an advanced form of General Systems Theory [GST], can provide a meta-theoretical concept through which subjective and objective observations can be brought together. Drawing from the notions of systems observation, design, perception, and participation, TCST outlines an epistemological theory that is capable of combining an exo-view and an endo-view of the world. In this way TCST naturally has to leave behind the realm of classical logic and non-contradictory systems description, hereby accepting and

---

1. First published on 24 May 2007 in the online journal, *Global Spiral* (8/2) available at http://www.metanexus.net/Magazine/tabid/68/id/10040/Default.aspx.

incorporating paradoxical experiences of the natural world into a new type of systems theory and systems knowledge.

TCST joins the knowing subject and the object under observation in conceiving of the observer/designer as an access system, whose properties (in similarity and difference) are likewise found in the system of observation. Finally, TCST arrives at an understanding of natural processes and occurrences in which freedom (the principal property of the human access system that is imperative for understanding its own self within the natural world) can be introduced to an understanding of the essence or self of the world. This view, however, will not confuse the natural world with full moral consciousness, but will suggest that free natural processes stand in an essential relationship to human persons, thereby obtaining meaning beyond that of being simply designed, or of occurring by chance.

It is at this point where science and religion do not have to continue to argue which interpretation of nature obtains more plausibility, but can jointly pursue the search for nature's true purpose.

INTRODUCTION

WITHIN THE CURRENT SCIENCE and theology/religion dialogue, the debate that arguably is carried out with most vehemence pertains to the question whether nature's discernible processes are caused by an intelligent being—that we (i.e., the faithful) call God[2]—or by the presumed self-generated processes of natural evolution.[3] This paper will argue that this, perhaps more adequately termed, "finality vs. contingency"[4] controversy concerning the motion of the natural world, is, at its root, not bound to the specific doctrines of theology or the sciences, but more generally related to the question of how these two disciplines become aware of the natural world. Thus, at heart, the author believes that at the

---

2. Aquinas, *Summa Theologiae*, Ia q.2 a.3 co.

3. Among many others this debate is kept alive by Cardinal Christoph Schönborn, "Finding Design in Nature," NYT, July 7, 2005, <available from: http://www.millerandlevine.com/km/evol/catholic/schonborn-NYTimes.html>; "The Designs of Science," Copyright (c) 2006 First Things (January 2006).<available from: http://www.firstthings.com/article.php3?id_article=7 >; *Ziel oder Zufall? Schöpfung und Evolution aus der Sicht eines vernünftigen Glaubens*.

4. Ordinarily this controversy is presented with referring to the notions of *intelligent design* and *random evolution*.

bottom of this debate lies an epistemological problem, i.e., how to understand and interpret the reality that nature exists in motion, develops, and ostensibly evolves.

NATURE AS IRRATIONAL OBJECT

Taking St. Thomas' *Summa Theologiae*, and here more specifically his five ways of the proof of God's existence (I^a q.2 a.3 co.) as an example for demonstrating the argument behind a generally accepted epistemology of nature,[5] it first must be noticed that St. Thomas, in all five arguments, proceeds from sense experience (Certum est enim, et sensu constat, aliqua moveri in hoc mundo). Finally in the teleological line of reasoning of the fifth way he introduces what arguably turned into the enduring caesura between subject and object of knowledge. Thomas labels all natural objects as *non habent cognitionem*, thus deducing that all things *operantur propter finem*, must be directed by a being *cognoscente et intelligente*—"and this being we call God."

Reasoning in this manner, Thomas prepared the ground for the single epistemological premise[6] that theology and the sciences (most of the time presumably unconsciously) continue to employ in arguing and verifying their respective interpretations of the processes of and within nature. In one of his last articles on this subject, Cardinal Christoph Schönborn cites Thomas' *Commentaria in octo libros Physicorum* in order to demonstrate nature as *ars divinae* or, in simple terms, God bestowing finality to the natural world. Thus, Schönborn—representative for today's employ of scholastic theology—concludes that as one cannot perceive of nature as rational, nature's *intelligent* behavior must have an external cause, i.e., an intelligent being.[7]

In the dawn of the Enlightenment precluding this admittedly speculative assurance of nature's purpose given by God, Renè Descartes' skeptical reflection on human knowing, likewise kept the *res cogitans* apart from the natural world or *res extensa*. While it can reasonably be assumed that natural objects exist, Descartes insists

---

5. Likewise in the *Summa*, a.I, q.2, Thomas speaks of *irrationalis natura* and *irrationalis creatura* vis-à-vis the rational human person.

6. Perhaps is could likewise be argued that the emergence of the opposition of strict reason versus empathy and suffering is linked to the birth of tragedy and the coming of rationality found in Euripides and Socrates. Cf. Nietzsche, *Die Geburt*, 1872.

7. Schönborn, "Fides, Ratio, Scientia. Zur Evolutionismusdebatte," 6.

that they must not to be confused with the person's knowing essence.[8] Ostensibly based on this accepted ground of cognition, Immanuel Kant develops a more detailed account of the possibility of knowing nature in his *Critique of Judgment*.[9] Kant's epistemological approach can be seen as representative of any contemporary approach to the study of nature that does not ultimately resort to the data of revelation and the postulates of faith.

Paragraph 75 of the *Critique of Judgment* reads: "We are in fact indispensably obliged to ascribe the concept of design to nature if we wish to investigate it, though only in its organized products, by continuous observation." The concept of design, therefore, is the necessary maxim of our use of reason. Further developing this maxim, Kant concludes that: "[Since] we do not, properly speaking, *observe* the purposes in nature as designed, but only in our reflection upon its products *think* this concept as a guiding thread for our Judgment, they are not given to us through the Object." The philosopher concludes at the end of the same paragraph: "We cannot therefore judge objectively, either affirmatively or negatively, concerning the proposition: 'Does a Being acting according to design lie at the basis of what we rightly call natural purposes, as the cause of the world (and consequently as its author)?'"

Thus, Kant—like Descartes, St. Thomas, and present-day students of theology and of the natural sciences—affirms the absolute otherness of nature as an object, neither bestowed with intelligence nor cognition, and thus incapable of giving or revealing to the knowing subject true knowledge of itself.

Based on the aforementioned assessment that both theology and the sciences categorically separate the knowing subject (i.e., the human person) from the object of knowledge (i.e., nature), one can indeed not find any other solution to the observation of purposiveness in nature than to postulate either (a) its being caused by God, or (b) its being grounded in nature itself and its laws, whether such are considered random or determinate.[10] Thus, in the opinion of this author, the problem of nature's finality or contingency has ultimately ended up in aporia in which any third possibility—though often proposed—has not yet found

---

8. cf. Descartes, *Meditations* VI, 9.
9. Available from: http://oll.libertyfund.org/Home3/Book.php?recordID=0318.
10. *Catholic Encyclopedia*, "nature."

a legitimate and commonly recognized place.[11] Any forged remedies to this problem, like simply equating nature's laws and evolution to God's continuing presence in nature, will legitimately earn stern criticism from scientists[12] who—in the light of Kant—demand their observations to stay clear of religious metaphors, as well as from religious fundamentalism, strictly adhering to a spirit-nature dualism.

Although throughout history the strict separation of person and nature was challenged by non-dualist approaches (archaic myths, mysticism, philosophy, and theology), the established domains of religious and scientific knowing unswervingly maintain this distinction, either because of reasons of doctrine,[13] or on account of the integrity and proposed objectivity of the respective discipline.[14] Thus in the end one perhaps has to agree with Ludwig Wittgenstein who maintained that if one is confronted with two rival worldviews that at heart are incompatible, one has to choose either one or the other, much like in the act of conversion toward a conviction that cannot be challenged on the grounds of reason.[15]

---

11. Karl Rahner wrote more than twenty years ago, "Die Schultheologie macht sich ja wenig Gedanken darüber, daß die Materie vom Ursprung und Ziel her doch sehr ‚geistig' sein muß, wenn ihr Schöpfer absoluter Geist ist und gar nicht Ursache von etwas sein kann, das schlechthin geistlos ist." Rawer and Karl Rahner, "Weltall-Erde-Mensch," 48.

12. Cf. for example Miller, "The cardinal's big mistake."

13. Quoting from the *Catholic Encyclopedia*, "creation": "If the universe were 'informed' by a principle of life, there would not be that essential difference between inanimate and animate bodies which both science and philosophy establish; inanimate bodies would manifest signs of life, such as spontaneous and immanent activity, organs, etc. The materialistic principle, 'No matter without force, no force without matter' (Büchner), though, with some obvious qualification, true as to its first part, is untrue as to its second. Force is the proximate principle of action, and may be or not be, but it is not of necessity conjoined with matter. The principle of action in man is not intrinsically dependent on matter."

14. "When the French physicist Pierre Simon de Laplace explained his theory of the universe to Napoleon, Napoleon is said to have asked, "Where does God fit into your theory?" to which Laplace replied, "I have no need of that hypothesis." Taken from Schick, "Can Science Prove?"

15. "Where two principles really do meet which cannot be reconciled with one another, then each man declares the other a fool or heretic." "At the end of reason comes persuasion. (Think of what happens when missionaries convert natives)." Wittgenstein, *On Certainty*, 612.

Then again, both theology and the sciences, while initially resistant to any blunt deviation from their doctrinal viewpoints,[16] have come to realize that the ultimate conclusions of their epistemological hypotheses on nature remain wanting. Theology does recognize that nature's teleological development to a great extent—if not entirely—follows the laws of mutation and selection,[17] while on the other hand serious scientists do tend to leave behind Laplace's belief that the future of the universe can completely be determined by scientific forecast.[18]

This paper proposes that while one cannot solve the above outlined dilemma within the accepted classical presuppositions of knowing, hereby referring to standard Aristotelian logic operating on the principle of non-contradiction,[19] an innovative approach to the subject—albeit offered on scientific grounds—will present a remedy. Theology,[20] and possibly the social sciences too, have already definitively answered the analogous problem of the finality of the human person by introducing the notion of moral freedom.[21] By possessing the faculties of reason and will, human persons experience themselves as free, whilst at the same time being bound to the laws of nature. Thus purposeful (i.e., meaningful) human behavior *can* be understood between two poles: freedom and determinateness.[22]

---

16. Cf. "Montium Concerning the Writings of Teilhard de Chardin."

17. John Paul II wrote: "In his encyclical *Humani Generis* (1950), my predecessor Pius XII has already affirmed that there is no conflict between evolution and the doctrine of the faith regarding man and his vocation, provided that we do not lose sight of certain fixed points ... Today, more than a half-century after the appearance of that encyclical, some new findings lead us toward the recognition of evolution as more than an hypothesis. In fact it is remarkable that this theory has had progressively greater influence on the spirit of researchers, following a series of discoveries in different scholarly disciplines. The convergence in the results of these independent studies—which was neither planned nor sought—constitutes in itself a significant argument in favor of the theory." Online: http://www.ewtn.com/library/PAPALDOC/JP961022.HTM.

18. Hawkins, "Does God play Dice?"

19. Aristotle, *Metaphysics*, Γ 3–4.

20. *Summa*, q.I, a.3. (I^a-IIae q.1 a.3 co) "Et utroque modo actus humani, sive considerentur per modum actionum, sive per modum passionum, a fine speciem sortiuntur. Utroque enim modo possunt considerari actus humani, eo quod homo movet seipsum, et movetur a seipso."

21. Hauser, *Moral Minds*.

22. Smit, "Free Will."

In this manner, this chapter proposes that a contribution to the question of finality vs. contingency involves, and is in need of, two major advances: First, the proposal and presentation of a new epistemological approach to nature that while proceeding from the experience of nature as purposefully behaving object(s), seeks to understand this experience within the analogical experience of the knowing subject. Trans-Classical Systems Theory outlined by the late Austrian Physicist Alfred Locker[23] will provide the theoretical foundation of this approach that can be regarded the endeavor to reach an inside or *endo-view* of nature.

Second, the chapter will attempt to bring the center of this experience (i.e., the notion of human freedom) to bear on an understanding of nature's behavior as *meaningful*. With regard to nature, the concept of "meaning" (*Bedeutung*) can be developed in view of its relationship to the subject, hereby aiming at a convergence of the viewpoints of theology and science. Natural processes understood as meaningful will reestablish the lost relationship of subject and object, and in this way invite a new ethical reading of nature.

SYSTEMS THEORY

General Systems Theory

The origins of General Systems Theory [GST] can be traced back to the Austrian biologist and polymath Ludwig von Bertalanffy (1901–1972),[24] widely considered the founding father of synthetic biology, whose aim for the most part was to oppose the then-dominant deterministic understanding of biological organisms.[25] The concept of system—according to Kant the "idea of collective nature . . . in accordance with the rule of purposes, to which Idea all the mechanism of nature must be subordinated according to principles of Reason (at least in order to investigate natural phenomena therein)"[26]—offered von Bertalanffy the tool to argue that the sciences must understand nature and biological organisms not only

---

23. Locker, "Glimpses of Truth," 103–5; "Obituary Alfred Locker," 571–75; Jung, "Alfred Locker: An Obituary," 1665–66; Pichler, "Alfred Locker Im Gedenken," 59.

24. Bertalanffy, *Theoretische Biologie*; Bertalanffy, *Robots, Men and Minds*; Bertalanffy, *General Systems Theory*; Bertalanffy, *The Organismic Psychology*; Bertalanffy, *Perspectives on General Systems Theory*; Bertalanffy, *A Systems View of Men*; and Bertalanffy, *Perspectives on General Systems Theory*. See also Davidson, *Qu∃rDenken! Leben*.

25. Taux, "Die Verwendung," 83–88.

26. Kant, *Critique of Judgment*, 67.

in view of the mechanical function of their parts, but as a system that is constituted by its distinction from an environment and the definitive relationship of its properties to 1) one another, 2) the entire outside world, and, 3) foremost, to the human person that is conceptually formulating and thus theoretically designing the system.[27]

GST established that systems, though initially appearing in difference to their observer must likewise be seen as 1) *subject*-analogs, and 2) *substance*-analogs to their designer. Only by viewing a system in relationship to its designer (i.e., an observer involved in the conceptual recognition of the system), can its *Ganzheit* or wholeness be perceived. The true nature of a system—its *totality* or theoretical *self* (i.e., its *true*, in opposition to its *real*, nature)—is more than simply the numerical sum of its individual parts.[28] In this way von Bertalanffy not only presented a potent alternative to any apparent reductionism in the sciences, but he likewise opened up the natural sciences en route to acknowledge and dialogue with the humanities.

In sum, GST must rightly be acknowledged as the foundational theory bringing together scientific observations concerned with the object (G from the German *Gegenstand*) under investigation, and their conceptual presuppositions (V from the German *Voraussetzungen*). In this way GST opened up the search for a viable systems' view or theory that can conceptually combine the objective proprieties of a system with the systems theory originating from its designer. GST thus sought to move general systems description from an *ortho*-level of classical description (G-level) to a *meta*-level (G+V-level) from where a system can be perceived in its entirety. Von Bertalanffy's work, however, met an impasse as GST (pursued by practitioners drawing from Bertalanffy's ideas) continued to rely upon classical concepts of systems description. In this way ruling out any systems depiction not falling under the perceptions of classical logic, GST, while rightly being acknowledged as a necessary

---

27. When Ludwig von Bertalanffy [the founding father of systems theory] spoke of Allgemeine Systemtheorie it was consistent with his view that he was proposing a new perspective, a new way of doing science. It was not directly consistent, however, with an interpretation often presented concerning "general system theory," to wit, that it is a (scientific) "theory of general systems." See Lazlo, foreword to *Perspectives on General Systems Theory*, 30. This was a collection of essays gathered together and published two years after his death in 1972.

28. Bertalanffy, "Wandlung des Biologischen Denkens," 352.

force in new approaches to the general study of nature, remained widely underappreciated.[29]

Trans-Classical Systems Theory

One of the last personal friends and students of von Bertalanffy, the Austrian physicist Alfred Locker (1922–2005) believed that GST's strengths could be retrieved if it leaves behind the constraints of classical scientific thinking.[30] According to Locker, this could be achieved on the one hand by more clearly outlining the role of the observer of a system in view of the system's perceivable nature, and, on the other hand, by introducing non-classical thinking to the propositions flowing from the *meta*-level of systems observation.[31] Alfred Locker coined this advancement of GST Trans-Classical Systems Theory [TCST] first introducing this concept in the late eighties.[32]

From Systems Observer to Participant

To begin with, TCST fully recognizes the limitations of any systems view derived on the basis of a detached systems *observer*. While an allegedly indifferent observer position naively promises objective insights about the system's nature, it in fact violates the *self* of the system by 1) reducing it to an object[33] that 2) is controlled by the very concepts of the observer uses to define it.[34] In addition, the hereby-assumed position of system-*allology* (i.e., complete otherness to the system) will erroneously suggest that there are neither substantial, nor essential links of the system to its observer (i.e., the human person). Such a systems position, that can be deemed an *exo*-view of it, is always in danger of resulting in utter indifference and perhaps irresponsibility toward the system under observation.[35]

---

29. Locker, "Horizontale und vertikale Relationalität."
30. Locker, "On The Ontological Foundations," 537–71.
31. Locker, "The Autological Foundation," 1–11.
32. Cf. Locker, "Recent Approach to Transclassical Systems-Theory," 11–16; Locker, "The Present Status of Classical Systems Theory," 8–16.
33. Locker, "Der Mensch," 35–38.
34. Locker, "[*il y a*] or The Withdrawal to the Portal of Being," 68–87.
35. Locker, "The Healing of Mankind's Predicaments," 131–52.

Alongside GST, TCST asserts that in reality any systems observer at the same time may be considered its designer. The process of design flows from the act of cognition in which various modes of understanding are introduced to the system. Hereby the designer formulates and explicates a systems theory according to which the system is conceptually constructed. This systems theory not only becomes the link between the designer and the system, but also establishes the system as *analog* to its designer in a twofold way; as *subject*-analog, the designer's self is the systemic basis of his design. As *substance*-analog the essence of the designer, here referring to the understanding of the self, is part of the theory used to comprehend and interpret the system. Thus TCST will conclude that even a system that is perceived in absolute otherness to its designer (e.g., nature), in order to imply this otherness must contain properties *analogous* to the designer.[36]

In establishing the relationship of the designer to the system, the limits of standard approaches to systems observation become apparent.[37] The true designer more and more recognizes the need to likewise *perceive* the system by way of intuition and imagination.[38] This recognition is based on the fact that a designer cannot anymore claim to remain outside the system at all moments of its perception. As systems-analog the designer permeates the conceptual systems border hereby assuming the role of a functional part of the design. This inside, or *endo*-view of the system calls for an altogether new theoretical approach to it.[39] TCST assumes that in an *endo*-view the system reveals its essence to the perceiver who is recognized as an integral part of it. However, this donation of the systems *being*[40] results in images that cannot be understood with the tools of classical epistemology. The perceiver has to resort to imagination drawing from inner experiences, similar to those found within the system. A systems perception more and more yields insights that allow outlining a systems image that, while grounded in *reality* (i.e., observation and design), describes its *totality*.

36. Locker, "Horizontale und vertikale Relationalität des Menschen."
37. Locker, "'Grenzen des Wissens,'" 48–51.
38. Locker, "Der Mensch," 34–42.
39. Cf. Rössler, "Endophysics," 154–62.
40 Cf. Here referring to the work of Jean-Luc Marion in Locker, "Systems Theory and the Conundrum," 297–317.

Within the attempt of formulating an increasingly holistic systems theory, the systems perceiver finally comes to realize that different positions assumed in view of the system not only generate different images of it, but also affect the notion of the *self* or essence of the system. At this moment, the perceiver recognizes his essential participation in the system and its influence on perceiving the notion of the system's *self*. This insight proves itself of crucial importance to current debates in the field of systems theory.[41]

If, as TCST holds, the *self* of the system is intrinsically linked to the self of the designer, than the system's *self* is a concept of lower order. Thus TCST remains hesitant to fully subscribe to concepts like self-reference or auto-*poiesis* as understood apart from the systems designer.[42] Following this premise, notions like Artificial Intelligence designs are equally obscure. If intelligence refers to the process of design, hereby conceiving of the theory of the system in view of the self of the designer, no engineered system can fully obtain or even create such self.[43] Equally unattainable would be the idea that biological systems gradually develop this self or any form of consciousness of it.[44]

Perhaps the real contribution of TCST can be found in arguing that in order to obtain a view of a system as whole (*Ganzheit, Gestalt, totality*) all four systems positions and views have to be taken together in forming an epistemological *access system* to any system in concern.[45] Within the access system, all four system positions are taken in an ever changing way in which observation–design–perception–participation form a dynamic circle of knowing. Any insight obtained from one systems position immediately seeks yet another view, leading to a new insight that, in turn, suggests assuming another systems position.

It is at this point where TCST claims to advance the epistemological status of the study of nature as outlined in the introduction of this paper. With regard to Thomas and scholastic theology, TCST, on the basis of viewing nature and its access systems as analogs, believes in the

---

41. Locker, "Der Mensch: Nicht unbeteiligter Zuschauer," 34–42.
42. Locker, "Metatheoretical Presuppositions," 209–33.
43. Locker, "A.I. and Ethics."
44. Locker, "Selbstentstehung," 33–69. Also Spämann, "Sein und Gewordensein," 73–92.
45. The notion of access system is further outlined in Locker, *Conundrum*, 300.

possibility of arriving at a true inside view of the world.[46] This inside view not only epistemologically intertwines the knowing subject and the objects of its cognition,[47] but likewise assumes that nature itself exudes its purposiveness and meaning to the human mind. In this sense TCST surpasses Kant's line of reasoning with regard to nature's purposes and its causes, however only in view of their analogical occurrence in the access system (i.e., the knowing human person). Thus TCST truly seeks to become a theoretical bridge between science, philosophy, and theology.

Toward a Non-Classical Systems Theory: Introducing and Solving Paradoxes

On the basis of the aforementioned proposal for a new epistemological entry point to the understanding of nature, TCST recognizes the need for an altogether new systems theory. Then again, this theory cannot simply remain on the level of non-contradictory logic, but has to acknowledge and embrace those paradoxes that inevitably show themselves when one and the same system is seen from different *exo* and *endo* systems positions.[48] In this way, however, TCST likewise overcomes what apparently already seems to be a paradox, i.e., that science and religion, both drawing from experience, postulate mutually excluding maxims. Thus, TCST, while initially appearing to conflict with established theories of knowledge, ultimately hopes to provide for the convergence of worldviews.[49]

Freedom and Meaning in Nature—A Trans-Classical View

As a consequence of understanding nature from the outside, theology decries that science simply dismisses the experience of *free will* as delusion.[50] However, today's scientific understanding of objectivity, dependency, and even randomness[51]—all found in nature—allows for the conclusion that whereas "not being a feature of nature, freedom is

---

46. Locker, "Systems-Theoretical Considerations."
47. Locker, "Kybernetik und Systemtheorie," 23–43.
48. Locker, "Reviving Paradoxes." Cf. Schöppe, *Theroieparadox*, and here especially § 3.3 "Die Beobachtung des Beobachters," 240–55.
49. Locker, "Recent Approach to Trans-Classial Systems Theory."
50. Cf. Townes, "Logic and Uncertainties," 53.
51. MacAndrew, "Life: Puppetry or Pageantry?"

carried out in nature."⁵² As follows, the established science-faith dialogue avows nature as condition of freedom, hereby to some extent considering *free process* within nature, while however continuing to reject any sense of moral freedom, and consequently rationality in nature and natural occurrences.⁵³

From this point of departure—that arguably reveals that advanced science likewise proceeds from a designer view (shown in the adoption of the concept of freedom to the description of nature)—TCST will further propose that if freedom is truly carried out in nature, nature it*self* accessed as system must be understood as containing properties related to a free *self* in analogy to the self assumed (cogito) by the knowing human person.⁵⁴

Similarly envisioning a joint inside and outside view of the world, the Jesuit theologian and paleontologist Pièrre Teilhard the Chardin (1881–1955) equally recognizes that the absolute concept of the view without (i.e., the outside) of things "completely breaks down with the human person, in whom the existence of a within can no longer be evaded, because it is object of a direct intuition and the substance of all knowledge."⁵⁵ Teilhard reaches the conclusion that the previous and perhaps up to now upheld distinction of the *determinate* without, and the *free* within—as irreducible and incommensurable—must be overcome. He masterfully argues against this dichotomy by recognizing the complementary forms of physical and spiritual energy of all that is, i.e., the cosmos, as bridging this gap that for centuries forcefully separated God from nature. Teilhard finally concludes that "spiritual perfection (or conscious 'centreity') and material synthesis (or complexity) are but the two aspects or connected parts of one and the same phenomenon."⁵⁶

Teilhard's insights draw from the genuine Christian experience that true moral freedom rests on spiritual freedom that cannot fully be attained by human merits, but is bestowed on the believer as Grace by virtue of Christ's sacrificial death on the cross, and received through baptism. This tradition with regard to "freedom in Christ"

52. Hattrup, "Freedom as Shadow Play."
53. Bereiter-Hahn, "Biologische," 31–57.
54. Locker, "On the Origin of Systems," 95–103.
55. Pièrre Theilhard de Chardin, *Human Phenomenon*, 55.
56. Ibid., 60.

first has been articulated by St. Paul (cf. Gal 2:4; 5:1, 13) in assuming that through the Resurrection "creation itself will be set free from its bondage to decay and obtain the glorious freedom (evleuqeri,a) of the children of God" (Rom 8:21). On the basis of Paul's assertion, the great Swiss theologian Karl Barth (1886–1968) rightly concludes in his famous commentary on Paul's Letter to the Romans: "if men are free, the world must be free also."[57]

Having postulated a free world, TCST must equally assume that freedom found within the world to a minimal degree implies moral freedom. Whereas this assumption certainly cannot lead to the affirmation of the postulates of neo-Platonism, i.e., that the whole world contains a rational soul, TCST would indeed agree with St. Thomas' understanding of souls, and of animals and plants.[58]

More important, however, is that such understanding is always linked to and dependent on the fact of seeing the systems designer as participant in, and even constituent of, the system in concern. In view of the notion of freedom, this means that free acts in nature only obtain authentic freedom (i.e., moral freedom) if they are seen in relation to a human person participating in the processes and actions of nature. Acknowledging this participation, then leads to the need to articulate the meaning of these actions, rather than to simply describe them as processes.

The Meaning of Nature

All systems can be studied in view of their behavior and goal(s). TCST, however, assumes that systems goals, neither can be exhaustively defined by laws and principles found to operate within the system or by its supposed or real origin or design (i.e., a teleo-*nomic* systems view), nor by its particular reaction to various events or influences (teleo-*zetic* systems view). Only a view of the system's *Ganzheit* (hereby joining observer and observation) reveals its capacity to set goals for it*self* in view of pursuing a particular aim or purpose.[59] This teleo-*genetic* systems view allows for recognizing freedom in nature as intrinsically linked to its goal. Then again, this goal might not simply be described as an end-point towards

---

57. Barth, *Epistle to the Romans*, 310.

58. At this point, this paper continues Fergus Kerr's assumption that one can conclude from a form of life to its essence. Cf. Kerr, *Theology after Wittgenstein*, 176–77.

59. Locker, "Predicaments," 131.

which the system moves, but as the system's capacity to behave meaningful in view of humankind. That means that in TCST nature, within the experience of human freedom, can be comprehended as a meaningfully acting system. The meaning of nature, however, eludes mere scientific description in favor of expressions of human experiences, like beauty and pleasure, or fear and suffering.

An example of how, in view of an ethics of animal experiments, this notion of "meaning" can apply to nature can likewise be found in the work of Alfred Locker. Remaining critical of the naïve conjectures of evolutionism (*Evolutionslehre*) throughout his entire life,[60] he explains in a lecture delivered in 1986 that the notions of *empirical* and *intelligible* freedom must be strictly separated.[61] While the latter refers to human actions, the former can, for example, be attributed to animals, who guided by instinct cannot truly be called free, i.e., being deemed as capable of reaching moral decisions.

Then again, eighteen years later, and on the basis of a refined account of TCST, Locker added to the above view the claim that from a *meta*-level of systems cognition, this apparent opposition between subject (human person) and object (animal) can be overcome by seeing all creatures in complementary unity to the human person.[62] Hereby forging a lasting unity of creation and all creatures, he explains that any creature seen in its *Gestalt* obtains a *self*. It is this *self* of an animal that is affected, and eventually suffers in the case of experiments. Alfred Locker concludes his thoughts stating:

> The differentiation between a sacrifice (spiritually surrendering for a greater good) and a victim (involuntarily subjected to sufferings) reveals that the experimental animal primarily belongs to the latter. But it can be elevated to the former when the full meaning of its suffering becomes obvious. The same holds true for "human testing," if, in contrast to the formidable atrocities, e.g., of concentration camps, the momentum of voluntariness is guaranteed, as pioneers of medical research frequently demonstrated by carrying out experiments on themselves.[63]

---

60. Locker, ed. *Biogenesis, evolution, homeostasis*; Locker, ed. *Evolution kritisch gesehen*; Locker, "'Evolution'—Ein faszinierender Ungedanke," 17–39; Locker, "Schöpfungs- und Evolutions-Problematik," 217–50.

61. Locker, "Evolutionstheorie," 20.

62. Locker, "Tierversuchs-Ethik und der 'Menschenversuch," 221–26.

63. Ibid., 221.

CONCLUSION

To sum up, a systems theoretical understanding of freedom in nature will neither be able to simply conclude that there is a single design to nature, nor that all its processes are random or determinate, but that the complementary aspects of freedom found in nature to human freedom, attest to its *meaningful* activities. In this way, not only nature and humanity can enter into a new existential relationship, but the human person will truly recognize its constitutive role in the purposiveness of nature. Thus it is hoped that the notion of *meaning in nature* might play a new reconciling role in the science-religion conversation.

10

Scripture and *Trans*-Science

Parables as "Systems" of the Kingdom[1]

MARKUS EKKEHARD LOCKER

ABSTRACT

At the end of nearly one hundred years of critical parable research, it still remains an enigma why some of the most interesting of these central stories of Jesus continue to confound their readers. e.g., why is a wedding guest, found without garment, thrown into darkness (Matt 20:13), and a servant who cannot multiply his single talent severely punished (Matt 25:29)? Whereas established scholarship seeks for ways to reduce these facts to moral metaphors, this paper believes that these apparent incongruities between text and Christian faith are caused by not sufficiently recognizing that parables are foremost speech systems. By studying the parables within systems-theoretical considerations, what appear to be theological obstacles can be understood as system-communications that point to necessary interaction and active communication with the system. Thus, this proposed joining of exegesis and systems-theory offers new and, hopefully, equally acceptable ways of reading the parables of Jesus.

---

1. Originally published as "Scripture and *Meta*-Science: Parables as 'Systems' of the Kingdom." *The Loyola Schools Review. School of Humanities* IV (2005) 59–83.

PREFACE

MODERN EXEGESIS EMBARKS ON its study of the parables of Jesus first and foremost by distinguishing their literary form.[2] In this way, these relatively short narratives are identified with speech units that are basically classified as a parable or similitude[3] and interpreted as an allegory or metaphor.[4] Whereas this exegetical modus operandi for a great number of parables leads to satisfactory results, it appears that some parables cannot find complete resolution in this way of interpretation.

The so-called "tragic" parables,[5] in particular, pose considerable interpretative problems for contemporary exegesis. Examples for these parables are the parable of the "Wedding Banquet" (Matt 22:1–14), the "Wise and Foolish Virgins" (Matt 25:1–13),[6] and the Talents (Matt 25:14–30), whose specific storylines, in one way or the other, are not compatible to the general worldview generated by Jesus' words and deeds.

In the parable of the Wedding Banquet, most exegetes assume that the wedding gown in Matt 22:12 refers to those good deeds or moral actions[7] that all members of the Matthean community—and therefore all

---

2. Among the forerunners to this approach, the most prominent is Rudolf Bultmann, *Die Geschichte der Synoptischen Tradition*, cf. ch. I.5 "Gleichnisse und Verwandtes," 179–222.

3. "The parable in the wider sense includes the similitude, but in a narrower sense is contrasted with it. The *similitude* tells of 'a typical situation or a typical or regular event,' the *parable* 'some interesting particular case.'" Linnemann, *Parables of Jesus*, 1.

4. Fundamental studies on the parables are: Crossan, *In Parables*; Jeremias, *Parables Of Jesus*; Via, *The Parables*; Jülicher, *Die Gleichnisreden Jesu*; Teil, *Die Gleichnisse Jesu im Allgemeinen*; Zweiter Teil, *Auslegung der Gleichnisreden*; Linnemann, *Gleichnisse Jesu*; Summaries on different approaches are found with Gowler, *What are they saying about the Parables*; Harnisch, *Gleichnisse Jesu*; Dschulnigg, "Positionen des Gleichnisverständnisses im 20," 335–51. Also Blomberg, "Interpreting the Parables," 51–78, as well as Arens, *Kommunikative Handlungen*; The Entrevernes Group, *Signs and Parables*; Aurelio, *Disclosures in den Gleichnissen Jesu*; Harnisch, *Die Gleichniserzählungen Jesu*; Harnisch, ed., *Die Neutestamentliche*; Meurer, *Die Gleichnisse Jesu als Metaphern*.

5. Via, *Parables*, 110–44.

6. At this point these three examples taken from Matthew's Gospel seem to be sufficient to illustrate the intended arguments. Extensive literature to them is found in, Hultgren, *Parables of Jesus*, 349–51; 178–79; 280–81.

7. "Modern interpreters have suggested that the wearing of such [wedding garments] signifies any of the following: (1) as with Irenaeus, good works: "a live lived in conformity with the Christian Laws," (B. Smith), "evidential works of righteousness" (Gundry), "the deeds of Christian discipleship" (Donahue), or similar expressions concerning righteousness or moral rectitude (Dawson; Hagner; Jones; Funk)." Hultgren mentions five possible allegorical interpretations for the wedding gown. Hultgren, *Parables*, 347.

Christians—are called to perform relentlessly. However, one should not overlook the fact that in this parable the king's servant invites people, both good and bad, from thoroughfares to the marriage feast (v. 10). Therefore, to conclude, with the majority of scholars, that the guest who is found by the king without a wedding garment represents a person lacking good deeds, is not immediately supported by the text. On the contrary, any reader of this parable will not fail to show sympathy for a man suffering a fate reminiscent of this guest.

As to the parable of the Wise and Foolish Virgins, John Paul Heil explains it as follows: "When the foolish maidens, at the time that their lamps are going out, ask their wise fellow maidens to give them some of their oil (v. 8), the wise maidens wisely recognize that there is not enough oil for all and direct their foolish fellow maidens to go rather to those who sell oils and buy some for themselves (v. 9)."[8]

In a similar manner Ulrich Luz writes:

> One must not ask if the oil indeed would not have been enough for sharing it (without a doubt the sharing of the oil would have been a lovely feature of this story). Equally one must not ask if there was actually no olive oil in the house of the bride. Also one should not understand the refusal of the wise virgins allegorically, e.g., as the impossibility to share deeds. The narrator rather chooses from several possibilities to carry on his story in view of the fact that he intends a tragic ending of his parable. The wise virgins, therefore, do not say no, because the marriage dance will last such a long time that all their oil will be needed, or because they are malicious, gloating or parsimonious, but simply because the story does go that way.[9]

The manifold explanations and instructions of the above exegetes aiming to suppress any factual interpretation of this parable—that there was no sharing of oil—cannot prevent the reader from getting the impression that the "wisdom" of the five wise virgins shows itself only in the realization that any sharing of oil with the five foolish ones would likewise prevent themselves from reaching their desired goal. This, however, represents a notion of "wisdom" that no exegete is able to reconcile with the words Jesus spoke earlier in the gospel: "For everyone who asks receives, and everyone who searches finds, and for everyone who knocks,

---

8. Carter and Heil, *Matthew's Parables*, 194.
9. Translated by the author from Luz, *Das Evangelium nach*, 477.

the door will be opened. Is there anyone among you who, if your child asks for bread, will give a stone?" (Matt 7:8f).

In a similar way, the well-known parable of the Talents (Matt 25:14–30) is replete with theological difficulties. With reference to an identification of the "man who went out on a journey" with Christ, reaching back into the early church,[10] Robert H. Gundry explains: "The portrait of the master as taking what does not belong to him should bother us no more than the comparison of Jesus' coming to that of 'a thief in the night' bothers us. The point of comparison does not have to do with thievery, but with unexpectedness. So also here, the portrait has to do with the forcefulness of Jesus' demand for good works, not with the ethics of taking what belongs to others."[11] Accepting this explanation in view of the literary form of *parable*, the final destiny of the servant who is simply afraid of losing the little that was entrusted to him (cf. vv. 21–23) posits a noticeable contrast to Jesus' invitation: "Take my yoke upon you, and learn from me; for I am gentle and humble in heart, and you will find rest for your souls. For my yoke is easy, and my burden is light" (Matt 11:29f.).

Thus far, the scholarship done on the parables generally gives the impression that the king in the parable of the wedding banquet judges on the basis of prejudices: the wise virgins act self-interestedly, and the man who went on a journey, not only distributes his talents in a biased way (Matt 25:15), but over and above, punishes the anxious servant excessively.[12]

PROPOSAL

This paper makes use of the fact that the linguistic approach employed in the abovementioned cases implies an interpretation of the parables on the formal basis of these texts viewed as a self-contained and autonomous communication system. This observation is reason enough to believe that more attention must be given to the fact that the parable narratives are foremost *communication*, or *speech systems* and, consequently, must be studied on the basis of systems-theoretical principles.[13]

10. Luz, *Matthäus*, 510.
11. Gundry, *Matthew*, 508.
12. Similar observations are made by Rohrbaugh, "A Peasant Reading," 35ff., and Herzog, *Parables as Subversive Speech*, 84ff.
13. For the foundational arguments of systems-theory and its further significance for linguistic systems, I am indebted to the study of Alfred Locker, "Über Entstehung und

By way of joining exegesis and systems-theory, it is hoped that these aforementioned difficulties in the interpretation of some parables may be addressed in a new way.[14]

It is especially important to emphasize that from a systems-theoretical point of view, language systems, like the parables, unfold their full meaning only if all their inherent pragmatic elements are 1) adequately described and articulated on all different levels of the system and 2) interpreted in relationship to the author and reader of the system.[15] In that way, not only the *speech event* of the system (what and how the parable narrates) is studied but also its own unique *communicative activity* (what the parable does) is sufficiently taken into account.

1) On the level of the speech event, the descriptive language of a speech system exposes the rules to which all communications of the system adhere. 2) The discourse or sum of the communicative actions of the system constitutes its border(s), and illustrates the difference of the system from its environment and other systems. 3) By way of its discourse, the system likewise interacts with other systems and its environment. The discourse of the system contains, for that reason, its formal objective and goal. On this level of speech, the system shows its potential to communicate and enter by way of its communicative actions into already existing systems, and hereby correcting them. Thereby, the whole pragmatic relevance and activity of the system in relationship to its environment and other systems in the same is brought to the fore. However, the full potential of a system's structural transformation is only then completely realized, when it actually modifies its environment and, at the same time, transforms and corrects parts of itself.[16]

Entwicklung formaler Systeme," 489–503, as well as "Meta-theoretische Voraussetzungen der Formalen und Empirischen Linguistik," 67–96.

14. Whereas authors have already pointed to the fact that "parables induce the listeners to make a decision after the mind of the narrator in a concrete historical situation" (Linnemann, *Parables*, 21), systems-theory will bring out that in reading the parables, the reader enters into this communicative systems of Jesus and actively participates in all inner communications of the system and its transformative activities.

15. Here following Krieger, *Einführung in die allgemeine Systemtheorie*, 132.

16. Ibid., 133.

FIGURE 10.1. The Communicative Activity of the Parable-System

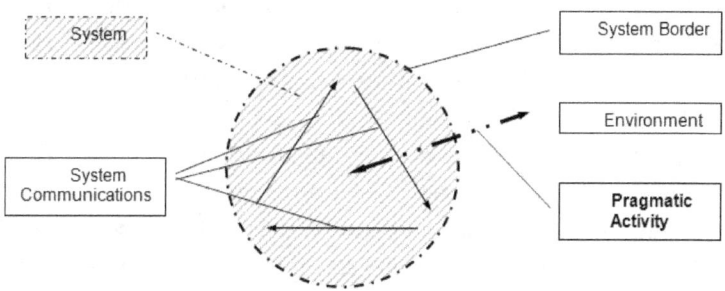

PARABLES AS SYSTEMS

The following paragraphs, in simplified form, will demonstrate in what way the parables of Jesus can be understood as scientific systems. The herein proposed systems-theory, however, cannot strictly be deemed *scientific* as, in its trans-classical form, it resembles a *meta-* or *trans-*scientific approach.[17]

The Nature of Systems

Systems,[18] in the most general sense, are designs that set themselves apart from a surrounding environment. Clearly distinguished from their environment, they constitute self-contained and self-referential units. However, as intentional reductions of the complexity of their outer world, systems are also authentic images of their environment. In this sense, systems exist and function by absorbing from and subsequently exuding to their immediate environment those properties that are essential for their interaction with it. This serves the purpose of constantly changing the environment through the system as to optimally correspond to it, in the end allowing for the possibility that the system and its environment enter into a balance. Spontaneous changes in the system itself, as well

---

17. Markus Locker, "System-Theorie," 16–22.

18. Fundamental works on systems-theory are: Bertalanffy, *Robots, Men and Minds*; "The System Concept in the Sciences of Man." For its trans-classical development, this study draws from works of Alfred Locker.

as in its environment, lead to disturbances and the necessity to newly establish the previously existing balance.[19]

Systems have an originator,[20] who designs the system by setting its boundaries (Jesus speaking the parables).[21] Devising a system, however, is possible only by assuming that every originator is himself part of yet another system (the world of Jesus). This system too has an environment, this time of unlimited complexities (the kingdom of God). In due course this system must have an unlimited originator (God). Thus, all communicative systems have two environments: first, an immediate environment of limited complexity {environment ②}, and second, and an absolute environment of unlimited complexity {environment ①} (see figure 10.2).[22]

According to Jesus' own words, his parables are an important part, if not the center, of his teaching (Mark 4:2). They contain the secret of the kingdom, i.e., the reign of God (Mark 4:11 par.) and are the key to enter into it. Astoundingly, while Jesus' parables speak of God's kingdom, they do not feature heavenly things but depict common and everyday occurrences. The kingdom of God is represented by commonplace events.

Parables are indeed systems; they are placed into this world originated by the spoken word of Jesus. The world here and now—theologically, this world after the fall—is their immediate environment, or environment ②. While in their descriptive nature, Jesus' parables are reductions of this world, they too, through their originator, represent an environment that is the world redeemed by Jesus, or their environment ①. In this way parables are authentic systems and images of the kingdom of God (cf. the introductory formula "the kingdom of God is like" [23]). On the basis of this analysis, one must not perceive or misunderstand the parables as the actual kingdom of God. As will be shown in the following discussion, the parables are better understood as systems, communicating with their environment as to introduce different possibilities to interact with it, and in this way to transform this world into the kingdom of God.

---

19. Most literature refers at this point to the examples of an air-conditioner, or a refrigerator.

20. Alfred Locker, "On the Origin of Systems," 95–103.

21. Krieger, *Systemtheorie*, 12–20.

22. Ibid., 15–16.

23. This point is especially developed by Jeremias, *Parables*, 101ff; Also Carson, "The ΟΜΟΙΟΣ Word-Group," 277–82.

FIGURE 10.2. The Environments of Parable-Systems

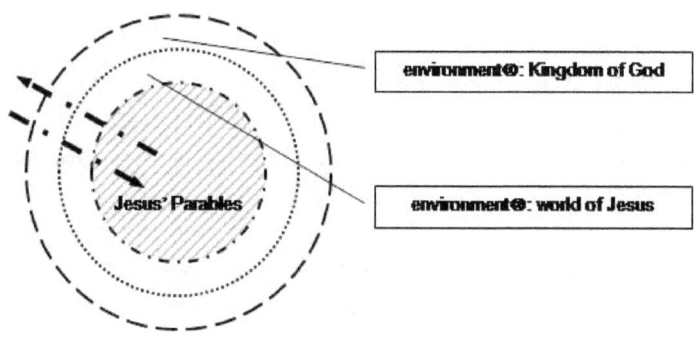

Since systems-theory assumes that any system environment shows genuine system-characteristics, the kingdom of God itself can be understood as a system.[24] But we must be careful not to identify the system of the kingdom of God with God. The system of the kingdom of God merely exhibits in what way (i.e., in what reductions) God becomes potentially visible in this world.[25]

The New Testament clearly shows the potential presence of God in this world in two different places. The Lukan evangelist portrays Jesus' seeing Satan falling like lightning from heaven (Luke 10:18) as an explanation as to why the demons' powers on earth were crushed (Luke 10:17ff.). This is the sign that the kingdom of God has been realized in this world (Luke 11:20).

At the end of the New Testament, in the book of Revelation, John sees that Michael cast Satan, the accuser, out of heaven (Rev 12:9ff). Thus, all Christians can follow Christ, the Lamb, to the kingdom-like city of the New Jerusalem.[26]

By speaking of occurrences and events of this world, parables represent the kingdom of God. They reduce the complexity of their worldly (in contrast to heavenly) environment by presenting selective actions taking place in this world. In this way, it is actually not the parable itself

---

24. Krieger, *Systemtheorie*, 13.

25. Ricoeur developed something similar in his hermeneutic of the metaphor. Ricoeur and Jüngel, *Metapher*, 27–34.

26. Markus Locker, "Das Buch der Offenbarung," 3–10; Locker, "Das Buch der Offenbarung im Verständnis der Sprachphilosophie," 115–29.

that represent the kingdom of God, but the actions and communications featured therein.

For example, that means that for the parable of the Workers in the Vineyard (Matt 20:1–16) one has to distinguish between the parable as a whole, and its communicative actions.[27] It certainly does not correspond to the nature of the kingdom of God that workers who labor for an entire day are paid inadequately, or to advocate communist-like customs. But it remains interesting that the provision and distribution of work for all waiting workers, as well as a just remuneration also for those who were only hired at the end of the day, point to a truly new and God-willed world.[28]

Classical Systems

At this point, it is especially important to bear in mind the difference between classical and trans-classical systems. Traditional systems-theoreticians believe that systems can completely represent and, as a result, hold all properties of their environment. Understood in this way, systems can be designed and studied individually and objectively (cf. figure 10.3).[29]

To understand the parables of Jesus as classical systems would inevitably result in the view that the communicative actions featured in them could be taken as objective norms of an objectively accessible world, naively believing that the kingdom of God could actually be objectively understood and erected here on earth by imitating or imposing these actions in and for this world. This view implicitly assumes that all actions of the system can be organized in such way that they do not oppose or contradict the system itself. Objective systems give the impression of being free of and free from any paradoxes.

Remembering the grumbling and indignant workers who worked all day in the vineyard (Matt 20:11f), it is not difficult to see that this systems-view has one major flaw. In an overly scientific manner, it assumes that systems can be designed separately and independently from

---

27. This, however does not mean that like for instance in the approaches of Propp, Greimas, and Güttgemans, one can study the communicative structures of the parables independent of their systemic nature. Cf. An exhaustive presentation of these approaches in Meurer, *Metaphern*, 71–142.

28. See Schottroff, "Human Solidarity," 129–47. Interesting approaches are also found in Lambrecht, *Out of the Treasure*, 85ff.

29. This, however, is the very presupposition that underlies all modern exegesis.

their designer or originator.[30] This, however, is impossible as any system observer has a blind spot (the position that he himself assumes designing and observing the system and, therefore, can not observe). Similar to a Heideggerian criticism of history, systems-theory must recognize that the designer/observer of a system is, at the same time, both an integral and ultimately, for its proper functioning, a necessary part of the same. This means that any system takes account of its observer as the already-transformed part of the environment that the system seeks to transform.[31] Within a trans-classical view the system articulates and allows for the paradox that is created by a designer/observer finding himself outside of and, at the same time, within the system.

Systems and Their Observers/Designers

The parables of Jesus must not be understood as objective worlds. They obtain their character from Jesus as their designer (D) and, therefore, must be seen as subject-analogs to him.[32] As a result, any observer (O) or reader of the parable is bound to assume first a similar analogical position to the system by accession of faith in Jesus. Through believing in Jesus the observer of the parables is transformed, too, into a subject-analog to them, and in this way enters the system, and changes position and function in view of this system.[33]

Reading the parables in faith places the reader into a constant dialogue with the system, in which on the one hand he tries to observe the system, and on the other hand, he enters and co-designs the system together with its originator (cf. figure 10.4). As designer, the reader of the parables is also invited to join and participate in their communicative actions in view of one's personal transformation and the world that is represented by them.

In theological terms, understanding Jesus' parables implies understanding Jesus, and therefore, believing in and heeding Jesus' call for repentance and *metanoia* (Mark 1:15 par.). Believing in his gospel is the precondition to join into the design of the parables, that is, to par-

30. Locker, Alfred, "Die Rolle des Beobachter-Subjekts," 89–101.

31. On this aspect of systems cf., A. Locker, "Horizontale und vertikale Relationalität des Menschen," 408–16.

32. A. Locker, "Healing of Mankind's Predicaments," 134.

33. A foundational study is A. Locker, "Schöpfungs- und Evolutions-Problematik," 217–49.

ticipate in the coming of his kingdom into a world that does not yet believe in Jesus and, therefore, still stands in contrasts to his parables.[34] Accordingly, seeing the vineyard's householder's fundamental commitment to provide work for all, shows that he has based his equal pay on his deliberate choice to realize equality and justice among his workers (Matt 20:13–16).

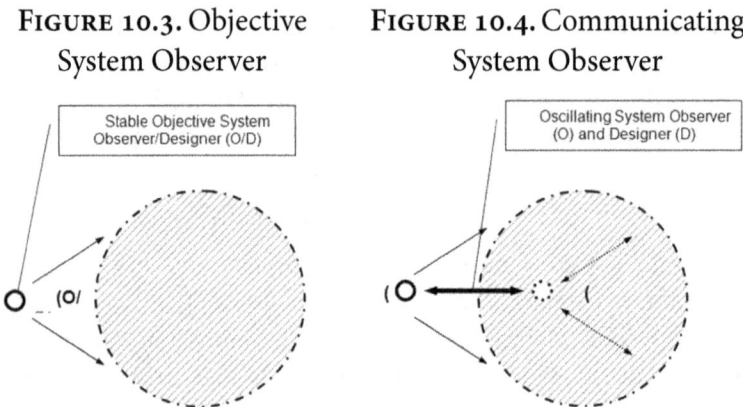

FIGURE 10.3. Objective System Observer

FIGURE 10.4. Communicating System Observer

Trans-classical Systems

Trans-classical systems are first of all characterized by positively accepting paradoxes within the system. These paradoxes do not weaken the system by rendering it unintelligible, but ultimately contribute to its strength.[35] Trans-classical systems-theory, not only recognizes the position of its designer and observer in the aforementioned way, but also acknowledges that systems, understood in classical terms, thus excluding paradoxes, are not able to fully transform their environment.

While on the basis of unavoidable system paradoxes, classical systems always separate two levels of observation: an ortho-level on which unity (U) and diversity (D) of the system oppose each other (e.g., the reality of salvation (U) and sin (D)), and a meta-level on which these oppositions are united (e.g., the redeemed human person U,(U,D)), trans-classical systems believe that these two levels can be combined, allowing

---

34. An interpretation of the book of Revelation in this line (cf. The formula *epoiēsen hēmas basileian* Offb 1,6, par 5,10) is presented in M. Locker, *Offenbarung in Ganzheitlich-Systemtheoretischer Deutung*, 4ff. Cf. also Locker and Clemens Sedmak, "The Language Game of Revelation," 241–62.

35. A. Locker, "Vorstoß zu einer transklassichen Sicht," 8–20.

the system to transcend its "classical" limitations, and in this way to truly transform its environment. Trans-classical-systems theory asserts that systems must incorporate, and not exclude paradoxes.³⁶

FIGURE 10.5. Trans-classical Systems

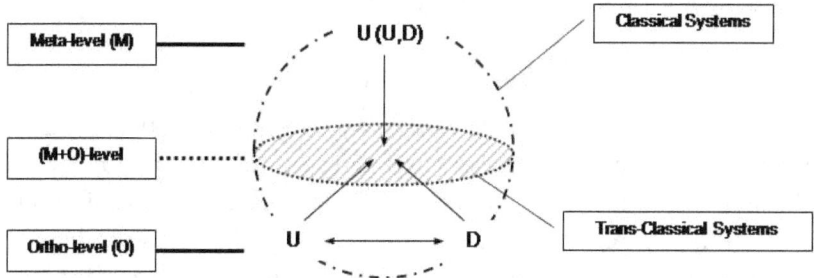

Parables must be understood as trans-classical systems. By transforming their observers, i.e., hearers, to accept faith in Jesus Christ, they begin to interact with their environment ②, in view of transforming this environment into a system like the kingdom of God—their environment ①.³⁷ Thus, the narrative level of the parables should only be identified with their ortho-level. Only if the observer is transformed through a parable does he move the text to its meta-level, and in this way understands it. Understanding the parables of Jesus, therefore, transforms the parable, its reader, and its immediate environment in view of God. Leaving the parables' meaning on their ortho-level, vis-à-vis an unattainable meta-level, reduces them to paradoxical narratives without real sense to be drawn from its communicative actions.³⁸

The good shepherd leaves his precious herd of sheep alone to find the one that went astray (Matt 18:12; Luke 15:4): the shrewd servant arbitrarily cancels the debts of his master (Luke 16:6f) and the house-

---

36. A good example here is the tendency of fundamentalism to see the human person at all times in the conflict of sin and holiness, instead of recognizing and addressing in a more fruitful way the paradox of *simul iustus et peccator* (Cf. Luther's lecture on Romans: Rom 4,7f . . . Beatus vir cui non imputabit Dominus peccatum)."

37. A. Locker, "Der Mensch," 34–42.

38. A. Locker, "System of the Un-Systematizable," 17–22.

holder of the vineyard orders his servant to pay all workers the same one talent, regardless of the length of their work (Matt 20:8).

It must, however, be assumed that the kingdom of God can be realized in no other way than in these paradoxes. While at first glance they render the parables unreasonable, these actions presuppose genuine faith and inspire the hearer and believer to completely enter into this world of faith, by acting according to it.[39]

The Goal of Systems

Trans-classical systems are teleogenic systems. Systems that include their designer have the capacity to set and seek goals in harmony with themselves and their designer. Teleogenic systems seek the realization of a goal that is inherent to them.[40] The general goal of Jesus' parables is to bring about the arrival of the kingdom of God through transforming their readers into realizing the communicative actions narrated by them. These actions delineate the discourse of the parables, and therefore, do not constitute descriptive propositions about the kingdom. In forming the border of the system, they distinguish it from its immediate environment ②, and likening it to its environment ①. Therefore, the ambit of the parables is not exclusively to be found within the systems' communications, but in the way these communications actually convey a worldview,[41] i.e., a new sense of this world that inadvertently allows for understanding the system itself.

Initially the border-discourse of a system illustrates possibilities of communication within the system itself allowing one to understand, alter, and apply its communicative actions in such a way that they truly realize all possibilities of the systems. At the same time the discourse-level of a system shows clearly what does not belong to the system, and what, therefore, must be excluded in its design and self-organization (cf. Rev 21:27 "but nothing unclean shall enter [the New Jerusalem]"). The parables of Jesus show and call for the possibility to love sinners, and in that way to rescue people who seem to be lost. They show that the possibility of justice, for example just wages, truly exists, and one can indeed waive the debts of one's debtors.

---

39. A. Locker, "Transclassical Turn of GST."
40. A. Locker, "Healing," 136.
41. Krieger, *Systemtheorie*, 133.

The discourse-level of the parables shows likewise how these systems can reach beyond themselves and transform their environment. Applying the parables' paradoxical actions to reorganize the systems themselves, at the same time, illustrates the various possibilities in which these actions could also transform other similar systems and, therefore, the immediate environment, i.e., the outer world of the system. Thus, a transformation of this world is unlikely to be a systemic transformation of the nature of the world, but a transformation of believers seeking to realize the transforming actions of the kingdom of God in all human systems contained in it.

Successful and Partially Successful Systems

In order to arrive at the solution this paper promised, one more important distinction among the parable systems must be pointed out. The introduction of this paper focused on those parables that, despite all the exegetical skills applied to them, continue to pose considerable interpretative difficulties. These parables give the appearance that, even in realizing their communicative possibilities, they do not completely transform this world into the kingdom of God. While the reader takes for granted that the parables of the Good Shepherd, the Workers in the Vineyard and the Shrewd Servant have happy and "comic" endings, the parables initially considered are only partial success stories. Leaving victims behind, these parables do not just depict a paradoxical but an outright "tragic" world.

Two considerations will lead to the final and most important insight of this paper:

1) If environments are represented and transformed by more than one system, then these systems and necessarily their designers/observers, do not simply interact with their environment one by one, but they interact with themselves while interacting with, and transforming their outer world (cf. figure 10.6)[42] Hereby systems that hold more transformative possibilities are supporting, and therefore, in some way equally transforming systems of less transformative nature.

2) Speech systems within the same environment, therefore, apply through their readers/hearers their selective communicative actions at the same

42. Ibid., 27ff.

time to themselves and to their environment. Therefore the discourse-level of the comic parables (Pc) equally demonstrates their possibility of transforming their neighboring tragic systems (Pt), even before these comic systems fully interact and communicate with their environment and readers/hearers.

FIGURE 10.6. Parable (P) Interaction and Transformation

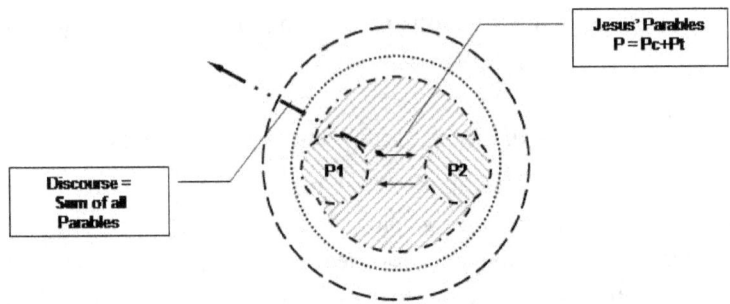

That means that an inclusive representation of the kingdom of God through the parables must start first, by allowing the border-discourse of the comic parables (e.g., Matt 18:12–14; Matt 20:1–16; Luke 16:1–20) to interact with the tragic parables (e.g., Matt 22:1–14; 25:1–13; 25:14–23). This course of action, then, will provide for a new way of understanding and interpreting the tragic parables. This, then, embodies a truly novel way of reading the parables of Jesus, viz. interpreting the tragic parables of Jesus in view of the communicative actions featured in the comic parables.

A SYSTEMS-THEORETICAL READING OF THE TRAGIC PARABLES OF JESUS

Jesus' parables are trans-classical systems that include their hearers and, in that way, invite them to participate in the transformation of this world. In this way, they presuppose the faith of their hearers transforming them in view of the kingdom of God. Jesus, i.e., God places the parables into this world to represent to their addressees the paradoxical nature of redeemed human beings living in this world, i.e., of the kingdom of God. However, the parables of Jesus are not in themselves the kingdom of God, but liken it by way of narrating communicative actions that open new possibilities of living according to God's kingdom to those who read and understand them in faith. Therefore, the kingdom of God is made visible and brought about through the parables of Jesus that on the one hand, interpret one another, and on the other hand interpret and change this world in view of a new way of life.

Comic Parables (Three Examples)

We shall here consider three examples of comic parables: The Good Shepherd (Matt 18:12–14), the Workers in the Vineyard (Matt 20:1–16), and the Shrewd Servant (Luke 16:1–20). These parables are characterized by communicative actions that challenge conventional responses to the common problems they feature. While it occurs all too often that one sheep runs off the flock, it is almost wrong to leave one's herd without protection to run after this one lost sheep. A flock without guard can become easy prey for predators or thieves. Likewise, no employer must pay everyone the same pay, but certainly favor a merit-based system; and canceling your boss's debts when in trouble, is outright theft.

The actions featured in these parables clearly presuppose that their agents base their understanding and reacting to the predicament that they face, in a truly unusual way, that is not at all the way actions are conceived of in this world (environment ② 2). One must believe that the parables agents are already transformed by faith in Jesus Christ, and that this faith constitutes an altogether new worldview out of which their decisions are made. This worldview exhibits traits of compassion, mercy, justice and love that have the power to turn looming tragedy into success. In the world in which these events happen, these actions of faith change existing rules to the effect of making a difference. Without the shepherd leaving his flock, the sheep would be doomed; without a just day's wage the workers last hired would never bring home enough money to provide for their families; and without his initiative, the shrewd servant would find himself without work. But despite looming and even expected tragedy, the sheep returns home (Matt 18:13; Luke 15:4), even casual part-time workers receive a fair salary (Matt 20:9), and a servant with one foot in prison is reinstated with honor (Luke 16:8).

Tragic Parables

Tragic parables (cf. Matt 22:1–14; 25:1–13; 25:14–23) have a wrong ending: they depict partial human failure. The wrongly dressed guest is literally thrown out (Matt 25:19), the foolish virgins are locked out (Matt 25:10) and even endure ridicule (Matt 25:12), and the anxious servant (Matt 25:25) who was not able to multiply his single talent, loses even this one to his fellow servants (Matt 25:28) before his master casts him out into the darkness (Matt 25:30).

The reader, and apparently the evangelist, seems to instantly know the reason for the success of some, and the failure of the other characters in these parables. Those who do not succeed in these stories were simply not able to meet the demands that the kingdom of God places on all believers. The average man on the street, as well as the foolish virgins, should have constantly awaited their invitation, as all Christians must vigilantly wait for the return of Christ. And the anxious servant should have overcome his worry and like his successful companions traded and multiplied his talent. We, too must not remain idle with our God-given talents, but multiply them.

Interpreting the Parable of the Talents[43]

A systems-theoretical reading of the parable of the talents demands more of its readers and believers than simply analyzing the reasons why the third servant acted wrongly. Could he really have overcome his anxiety, and risked his only talent?

A different interpretation unfolds in front of the reader's eyes if he is transformed from an onlooker who studies this parable in scientific terms into a designer of the parable, who by applying his own faith in Jesus Christ can transform this system into a comic parable. Thereby the reader actualizes his possibility to correct the system in a way that turns around the predictable fate of the third servant—due to his anxiety and perception of the master—without hereby endangering, but altogether enhancing the system's corrective function to its environment.

When the reader becomes part of the system, then he does not introduce himself to the system as hitherto foreign object, but he assumes the position of a system's inherent part or agent. Assuming for instance the role of the servant who received the five talents allows for two pos-

---

43. M. Locker, "Seeing Actions of Hope," 110–25.

sibly new observations and subsequent new actions: 1) seeing that even with only two talents, one does not have to be anxious, and with some degree of surety and certainty can trade and multiply them, and 2) realizing that one can redeem the worry of the third servant by conceiving of an action that, like those in the comic parables, has the potential to turn around looming tragedy, one can correct the existing system by giving one of one's five talents to the one servant with the single talent. In this way, the system is corrected, and all servants have the possibility to conform to the demands of the kingdom of God.

Assuming the role of one of the two other servants, likewise, opens up possibilities to correct the system. The reader, on the one hand appeals to the servant with the five talents to give up one talent for the one who has only a single one, and on the other hand seeks to overcome any kind of anxiety that deprives himself to trade and multiply his/her talents. Thus, any given possibility of corrective interaction within this parable opens up indefinite possibilities to bring about the kingdom of God on the basis of faith in Jesus Christ.

The Parables of the Wedding Feast and the Foolish Virgins

An interpretation of these parables that reaches beyond the already-accepted results of exegesis invites the readers to first creatively assume the role of the inviting servant (Matt 22:3) and the wise virgins (Matt 25:4). To know that the king judges his guests according to proper clothing brings any servant to the realization to provide for those when inviting people from the street. Knowing how foolish some persons can be, will likewise lead any reader who enters into the parable of the foolish virgins to bring an extra ration of oil on the journey. In both cases, these realizations offer new possibilities to transform these two parables from tragic into comic systems that in due course depict a world in which also foolish and unprepared people will find redemption.

CONCLUSION

A systems-theoretical interpretation of the parables of Jesus does not aim at reversing the results of existing parable research, but hopes to shed new light on their meaning and function in, and through a connection to their readers. This chapter concludes that on the basis of formal reasons, the parables of Jesus cannot be objectively understood as isolated speech units. Rather the parables are, like trans-classical system theory suggests, systems that include their interpreters as system elements necessary for realizing their full communicative success. Parables are systems that seek to bring about the kingdom of God by calling for faith and actions that represent God's reign here on earth in view of, and by imitating Jesus Christ, who is the originator and designer of the parables. In applying the actions depicted in the parables to the world into which the parables are spoken, the readers and believers originate a discourse that offers the possibility to live according to God's reign, transforming the earth, and at the same time personally accepting his salvation.

Bibliography

Abrams, Daniel. "New Study Tools from the Kabbalists of Today: Toward an Appreciation of the History and Role of Collectanea, Paraphrases and Graphic Representations in Kabbalistic Literature." *Journal des Études de la Cabale* 1 (1997). Online: http://jec2.chez.com/artabrams.htm.
Acland, C., and W. Buxton, editors. *Harold Innis in the New Century*. Montreal: McGill-Queen's University Press, 1999.
Alstyne, R. *The Rising American Empire*. New York: Norton, 1974.
Angyal, Andreas. "The Structure of Wholes." *Philosophy of Science* 6 (1939) 25–37.
Anselm. "Cur Deus homo." Forth Worth: RDMc, 2005.
Arendt, H. *On Revolution*. New York: Viking, 1963.
———. *The Human Condition*. Chicago: University of Chicago Press, 1958.
Arens, Edmund. *Kommunikative Handlungen. Die paradigmatische Bedeutung der Gleichnisse Jesu für eine Handlungstheorie*. Düsseldorf: Patmos, 1982.
Ashby, R. *Design for a Brain*. New York: Wiley, 1952.
———. *Introduction to Cybernetics*. London: Methuen,1956.
Assmann, A. "Was ist Weisheit?" In *Weisheit: Archäologie der literarischen Kommunikation III*, edited by Alida Assmann, 15–44. Munich: Fink, 1991.
Aurelio, Tullio. *Disclosures in den Gleichnissen Jesu. Eine Anwendung der disclosure-Theorie von I. T. Ramsey, der modernen Metaphorik und der Theorie der Sprechakte auf die Gleichnisse Jesu*. Regensburger Studien zur Theologie 8. Bern: Lang, 1977.
Auster, Paul, editor. *True Tales of American Life*. London: Faber & Faber, 2002.
Barth, Karl. *The Epistle to the Romans*. Translated by Edwyn C. Hoskyns. Oxford: Oxford University Press, 1968.
Bateson, G. *Mind and Nature*. New York: Bantam, 1979.
———. *Steps to an Ecology of Mind*. New York: Ballantine, 1972.
Belliger, A., and D. J. Krieger. *Ritualtheorien. Ein einführendes Handbuch*. Opladen: Westdeutscher, 1998.
Bellow, S. *Ravelstein*. New York: Viking, 2000.
Bennett, James G. *The Dramatic Universe*. Vol. 3. London: Hodder and Stoughton, 1966.
Benson, L. *Turner and Beard*. Westport: Greenwood, 1980.
Bereiter-Hahn, Jürgen. "Biologische Vorbedingung für die Ermöglichung freier Willensentscheidung." In *Gott, der Kosmos und die Freiheit. Biologie, Philosophie und Theologie im Gespräch*, edited by G. Fuchs and H. Kessler, 31–57. Würzburg: Echter, 1996.
Berger, P. *The Sacred Canopy*. Garden City, NY: Doubleday, 1967.
Bertalanffy, Ludwig von. *General System Theory*. Rev. ed. New York: Braziller, 1968.
———. *Perspectives on General System Theory*. New York: Braziller, 1975.

———. "The System Concept in the Sciences of Man." In *General System Theory: Foundations, Development, Application*, by Ludwig Bertlanffy, 186–200. New York: Braziller, 1968.

———. *Robots, Men and Minds: Psychology in the Modern World*. New York: Braziller, 1967.

———. "General System Theory—A Critical Review." *General Systems* 7 (1962) 1–20.

———. *The Organismic Psychology and Systems Theory*. Worcester, MA: Clark University Press, 1968.

———. "The History and Status of General Systems Theory." In *Trends in General Systems Theory*, edited by G. J. Klir, 251–69. New York: Wiley, 1972.

———. "Wandlung des Biologischen Denkens." *Neue Jahrbücher für Wissenschaft und Jugendbildung* (1934) 339–66.

———. *A Systems View of Men*. Boulder, CO: LaViolette, 1981.

———. *Perspectives on General Systems Theory: Scientific-Philosophical Studies*, edited by E. Taschdjian, 149–69. Braziller: New York: 1975.

———. *Robots, Men and Minds*. New York: Braziller, 1967.

———. *Theoretische Biologie*. Vols. 1–2. Berlin: Bornträger, 1932/1940.

Bevans, St. *Models of Contextual Theology*. Maryknoll, NY: Orbis, 1992.

Blank, P. "The Pacific: A Mediterranean in the Making?" *Geographical Review* 89 (1989) 265–77.

Blomberg, Craig L. "Interpreting the Parables of Jesus: Where Are We and Where do We Go from Here?" *Catholic Biblical Quarterly* 53 (1991) 50–78.

Bloom, A. *The Closing of the American Mind*. New York: Simon & Schuster, 1987.

Boscovich, R. J. "On Space and Time as They are Recognized by Us (in Latin)." Vienna, 1755. Reprinted in R. J. Boscovich, *Theoria philosophiae naturalis*. Vienna, 1758.

Brunnen, D. Pfannek am. *Die Entmutterung der Seele*. Leipzig: Engelsdorfer, 2004.

Buber, Martin. *Tales of the Hasidim: Later Masters*. New York: Schocken, 1961.

Buchanan, P. *The Death of the West: How Dying Populations and Immigrant Invasions Imperil Our Country and Civilization*. New York: St Martin's Griffin, 2002.

Bultmann, Rudolf. *Die Geschichte der Synoptischen Tradition*. Göttingen: Vandenhoeck & Ruprecht, 1931.

Bumpass, L. "Children's Experience of Marital Disruption." *American Journal of Sociology* 85 (1979) 49–65.

Bunge, Mario. *Method, Model and Matter*. Boston: Reidel, 1973.

Callon, M. "Some Elements in a Sociology of Translation: Domestication of the Scallops and Fishermen of St. Brieuc Bay." In *Power, Action and Belief*, edited by J. Law, 19–30. London: Routledge, 1986.

———. "Techno-Economic Networks and Irreversibility." In *A Sociology of Monsters*, edited by J. Law, 132–61. London: Routledge, 1991.

———. "The Sociology of an Actor-Network: The Case of the Electric Vehicle." In *Mapping the Dynamics of Science and Technology*, edited by M. Callon and J. Law, 19–34. London: MacMillan, 1986.

Carnochan, W. B. *The Battleground of the Curriculum: Liberal Education and American Experience*. Stanford, CA: Stanford University Press, 1993.

Carroll, J. *The Western Dreaming*. Sydney: Harper Collins, 2001.

———. *The Wreck of Western Culture*. Melbourne: Scribe, 2004.

Carson, D. A. "The ΟΜΟΙΟΣ Word-Group as Introduction to some Matthean Parables." *New Testament Studies* 31 (1985) 277–82.

Carter, Warren, and John Paul Heil. *Matthew's Parables*. The Catholic Biblical Quarterly Monograph Series 30. Washington, DC: Catholic Biblical Association of America, 1998.

Casteneda, Carlos. *Journey to Ixtlan*. New York: Touchstone, 1972.

Catholic Encyclopedia. "Nature." Online: http://www.newadvent.org/cathen/10715a.htm.

———. "creation." Online: http://www.newadvent.org/cathen/04470a.htm.

Chang, Chung-yuan. *Creativity and Taoism: A study of Chinese Philosophy, Art, & Poetry*. New York: Harper & Row, 1963.

Churchill, W. The Tragedy of Europe, 1946. Online: http://www.ellopos.net/politics/churchill-tragedy.htm.

Clarke, W. N. *Person, Being and Ecology*. Edited by R. Ibana. Quezon City: Ateneo de Manila University Office of Research & Publications, 1996.

Cochrane, C. *Christianity and Classical Culture*. New York: Oxford University Press, 1957.

Cohen, A. *The Tremendum: A Theological Interpretation of the Holocaust*. New York: Crossroad, 1981.

Connor, E. "Harold Innis." In *Key Thinkers for the Information Society*, edited by C. May, 87–108. London: Routledge, 2003.

Coulter, N. A., Jr., and Alfred Locker. "Recent Progress Towards a Theory of Teleogenic Systems." *Kybernetes* 5 (1976) 67–72.

Croly, H. *The Promise of American Life*. New York: Macmillan, 1909.

Crossan, John Dominic. *In Parables: The Challenge of the Historical Jesus*. New York: Harper & Row, 1973.

Dahl, R. *A Preface to Democratic Theory*. Chicago: University of Chicago Press, 1956.

Dan, Joseph. *The Ancient Jewish Mysticism*. Tel-Aviv: MOD, 1993.

Davidson, Mark. *Quer Denken! Leben und Werk Ludwig von Bertalanffys*. Frankfurt: Lang, 2005.

de Chardin, Theilhard (Pére). *The Human Phenomenon*. Translated by Bernard Will with an introduction by Sir Julian Huxley. New York: Harper & Row, 1975.

Descartes, R. *Meditations on the First Philosophy* (in Latin). Paris: Soly, 1641.

Dschulnigg, Peter. "Positionen des Gleichnisverständnisses im 20. Jahrhundert. Eine kurze Darstellung der fünf wichtigsten Positionen der Gleichnistheorie (Jülicher, Jeremias, Weder, Arens, Harnisch)." *Theologische Zeitschrift* 45 (1989) 335–51.

Lenneberg, E., editor. *Psychology and Biology of Language and Thought. Essays in Honor of Eric Lenneberg*. New York: Academic, 1978.

Eigen, M., and P. Schuster. *The Hypercycle*. Heidelberg: Springer, 1979.

Emerson, R. W. "English Traits." In *The Portable Emerson*, edited by C. Bode & M. Cowley, 395–53. New York: Viking Penguin, 1981.

———. "On the Relation of Man to the Globe." In *The Early Lectures of Ralph Waldo Emerson*, edited by S. Whicher and G. Spiller, 27–49. Cambridge: Harvard University Press, 1959.

The Entrevernes Group. *Signs and Parables. Semiotics and Gospel Texts*. Pittsburgh Theological Monograph Series 23. Pittsburgh: Pickwick, 1978.

Everett, H. "Relative-state Formulation of Quantum Mechanics." *Review of Modern Physics* 29 (1957) 454–62.

Fackenheim, E. *The Jewish Bible after the Holocaust*. Manchester: Manchester University Press, 1990.

Feher, F., and A. Heller. *Eastern Left, Western Left*. Cambridge: Polity, 1987.

Ferguson, N. *Colossus: The Price of America's Empire*. New York: Penguin, 2004.
Fine, L. editor. *Essential Papers on Kabbalah*. New York: New York University Press, 1995.
Fischer-Schreiber, Ingrid. *The Shambhala Dictionary of Taoism*. Translated by Werner Wünsche. Boston: Shambhala, 1996.
Foerster, H. von. "Cybernetics of Cybernetics." In *Communication and Control*, edited by K. Krippendorf, 5–8. New York: Gordon & Breach, 1979.
———. Foerster, Heinz von. *Observing Systems: Selected Papers of Heinz von Foerster*. Seaside, CA: Intersystems, 1981.
Franck, G. "Virtual Time: Can Subjective Time be Programmed?" In *Ars electronica Virtual Worlds*, edited by G. Rattinger et al., 57–81. Linz: Veritas, 1990.
Frank, Adolph. *The Kabbalah*. Translated by Dr. I. Sossnitz. New York: Kabbalah, 1926.
Fung, Yu-Lan. *A History of Chinese Philosophy*. Vol. II. Translated by Derk Bodde. Princeton: Princeton University Press, 1953.
Gadamer, H. G. *Truth and Method*. New York: Seabury, 1975.
Gans, Eric. *Originary Thinking. Elements of Generative Anthropology Structures*. Stanford, CA: Stanford University Press, 1993.
Gelwick, Richard. "The Polanyi-Tillich Dialogue of 1963: Polanyi's Search for A Post-Critical Logic in Science and in Theology." *The Polanyi Society* 22 (1995–96) 11–9.
Gianetti, C. *Ästhetik des Digitalen: Ein intermediärer Beitrag zu Wissenschaft, Medien- und Kunstsystemen*. Vienna: Springer, 2004.
Glanville, Ranulph. "A Note on Knowing." In *Theorie-Prozess-Selbstreferenz*, edited by O. Jahraus and N. Ort, 187–97. Konstanz: UVK, 2003.
———. "Learning from Locker." *Kybernetes* 35 (2006) 223–27.
Glaser, E., and H. Wellenreuther. *Bridging the Atlantic: The Question of American Exceptionalism in Perspective*. New York: Cambridge University Press, 2002.
Glasersfeld, E., von. *Radical Constructivism*. London: Falmer, 1995.
Gleason, P. *Keeping the Faith: American Catholicism Past and Present*. Notre Dame: University of Notre Dame Press, 1987.
Gödel, K. "An Example of a New Type of Cosmological Solutions of Ewinstein's Field Equations of Gravitation." *Review of Modern Physics* 21 (1949) 447–50.
Gordon, W. T. *Marshall McLuhan*. New York: Basic, 1997.
Gowler, David B. *What Are They Saying About the Parables?* New York: Paulist, 2000.
Guénon, René. *The Great Triad*. Translated by Peter Kingsley. Cambridge: Quinta Essentia, 1991.
Gundry, Robert H. *Matthew: A Commentary on His Handbook for a Mixed Church under Persecution*. Grand Rapids: Eerdmans, 1994.
Günther, Gotthard. "Die Philosophische Idee einer Nicht-Aristotelischen Logik." In *Beiträge zur Grundlegung einer operationsfähigen Dialektik*, edited by G. Günther. Vol. 1, 24–30. Hamburg: Felix Meiner, 1976.
Gutfeld, A. *American Exceptionalism*. Portland: Sussex Academic, 2002.
Gutiérrez, G. *On Job: God-Talk and the Suffering of the Innocent*. Maryknoll: Orbis, 1987.
Habermas, J. *Theory of Communicative Action*. Boston: Beacon, 1984.
Haken, H. *Information and Self-Organization*. New York: Springer, 1988.
———. *Synergetics: An Introduction*. New York: Springer, 1983.

Bibliography

Harnisch, Wolfgang, editor. *Die neutestamentliche Gleichnisforschung im Horizont von Hermeneutik und Literaturwissenschaft.* Wege der Forschung 575. Darmstadt: Wissenschaftliche Buchgesellschaft, 1982.

———. *Gleichnisse Jesu. Positionen der Auslegung von Adolf Jülicher bis zur Formgeschichte.* Wege der Forschung 366. Darmstadt: Wissenschaftliche Buchgesellschaft, 1982.

———. *Die Gleichniserzählungen Jesu. Eine Hermeneutische Einführung.* Göttingen: Vandenhoeck & Ruprecht, 1985.

Hartz, L. *The Founding of New Societies.* New York: Harcourt Brace, 1964.

———. *The Liberal Tradition in America.* New York: Harcourt Brace, 1955.

Hattrup, Dieter. "Freedom as Shadow Play of Chance and Necessity." E-conference proceedings: Continuity and Change 2006. Philadelphia, June 2006. Online: http://www.metanexus.net/conferences/pdf/conference2006/Hattrup.pdf

Hauser, Marc D. *Moral Minds: How Nature Designed Our Universal Norms of Right and Wrong.* New York: Harper Collins, 2006.

Havelock, E. *Harold A. Innis.* Toronto: Harold Innis Foundation, 1982.

———. *Preface to Plato.* Oxford: Blackwell, 1963.

———. *The Liberal Temper in Greek Politics.* New Haven: Yale University Press, 1957.

———. *The Literate Revolution in Greece and Its Cultural Consequences.* Princeton: Princeton University Press 1982.

Hawkins, Stephen. "Does God Play Dice?" No pages. Online: http://www.hawking.org.uk/lectures/dice.html.

Hegel, G. W. F. *The Philosophy of History.* New York, Dover, 1956.

Heidegger, M. *Basic Concepts.* Bloomington: Indiana University Press, 1988.

———. *Being and Time.* Oxford: Basil Blackwell, 1980.

Henderson, John B. *Correlative Cosmology in the Neo-Confucian Tradition. The Development and Decline of Chinese Cosmology.* New York: Columbia University Press, 1984.

Herzog, W. *Parables as Subversive Speech: Jesus as Pedagogue of the Oppressed.* Louisville: Westminster John Knox, 1994.

Heyer, P. *Communications and History.* New York: Greenwood, 1998.

Hietala, T. *Manifest Design: American Exceptionalism and Empire.* Ithaca, NY: Cornell University Press, 2003.

Hobsbawm, E. *The Age of Extremes: The Short Twentieth Century, 1914–1991.* London: Michael Joseph, 1994.

Hooff, J. A. R. A. M. van "A Comparative Approach to the Phylogeny of Laughter and Smiling." In *Non-Verbal Communication,* edited by R. A. Hinde, 209–41. Cambridge: Cambridge University Press, 1972.

Hultgren, Arland J. *The Parables of Jesus: A Commentary.* Grand Rapids: Eerdmans, 2000.

Hutchins, R. M. *The Higher Learning in America.* New Haven: Yale University Press, 1936.

Idel, Moshe. *Kabbalah: New Perspectives.* New Haven: Yale University Press, 1988.

Innis, H. *Empire and Communications.* Oxford: Clarendon, 1950.

———. *The Bias of Communication.* Toronto: University of Toronto Press, 1951.

———. *The Idea File of Harold Adams Innis.* Toronto: University of Toronto Press, 1980.

Jacobs, J. *Cities and the Wealth of Nations.* New York: Vintage, 1984.

———. *The Economy of Cities.* New York: Vintage, 1970.

Jenkins, R. *Churchill: A Biography.* London: Penguin, 2002.

Jeremias, Joachim. *The Parables of Jesus.* Translated by S. H. Hooke. Rev. Ed. New York: Scribner's, 1963.
John Paul II. "Message to the Pontifical Academy of Sciences on Evolution." Online: http://www.ewtn.com/library/PAPALDOC/JP961022.htm.
Johnson, C. *The Sorrows of Empire.* New York: Metropolitan, 2004.
Jülicher, Adolf. *Die Gleichnisreden Jesu, Zwei Teile in einem Band. Erster Teil, Die Gleichnisse Jesu im Allgemeinen. Zweiter Teil, Auslegung der Gleichnisreden der ersten drei Evangelien. Unveränderter reprographischer Nachdruck der Ausgabe Tübingen 1910.* Darmstadt: Wissenschaftliche Buchgesellschaft, 1969.
Jung, Richard. "Alfred Locker: An Obituary." *Kybernetes* 34 (2005) 1665–66.
Kant, Immanuel. *Critique of Judgment.* Translated by J. H. Bernard. London: Macmillan, 1914.
Kaplan, Aryeh. *Meditation and Kabbalah.* New York: Weiser, 1982.
———. *Bahir.* York Beach, ME: Weiser, 1979.
———. *Sefer Yeszirah: The Book of Creation.* New York: Weiser, 1990.
Kauffman, S. A. *The Origins of Order: Self-Organization and Selection in Evolution.* Oxford: Oxford University Press, 1993.
Kerr, Fergus. *Theology after Wittgenstein.* New York: Blackwell, 1988.
Kirk, R. *The Roots of American Order.* La Salle, IL: Open Court, 1974.
Körner, St. *Categorial Frameworks.* Oxford: Blackwell, 1970.
Krieger, David J. *Einführung in die allgemeine Systemtheorie.* Munich: UTB, 1996.
———. *Selbstreferenz und Subjektivität. Zur Konstruktion von Identität in der multikulturellen Gesellschaft.* Interedition. Megen: IKF, 1997.
Kriese, W., and O. E. Rössler. *Encouraging Lampsacus.* Rottenburg: Mauer Verlag, 2001.
Krings, H. "Sapientis est ordinare." In *Philosophie und Weisheit,* edited by W. Oelmüller, 161–65. Paderborn: Schoeningh, 1989.
Kristol, I. *Reflections of a Neo-Conservative.* New York: Basic Books, 1983.
Krohn, W. and G. Küppers, G., editors. *Emergenz: Die Entstehung von Ordnung, Organisation und Bedeutung.* Frankfurt: Suhrkamp, 1989.
Krohn, W., G. Küppers, and H. Novotny, editors. *Self-organisation: Portrait of a Scientific Revolution.* Boston: Reidel, 1990.
Lambrecht, Jan. *Out of the Treasure: The Parables in the Gospel of Matthew.* LTPM 10. Louvain: Peeters, 1991.
Latour, B. *Science in Action: How to Follow Engineers in Society.* Milton Keynes, UK: Open University Press, 1987.
———. "Drawing Things Together." In *Prepresentation in Scientific Practice,* edited by M. Lynch and S. Wollgar, 19–68. Cambridge: MIT, 1990.
———. "The Powers of Association." In *Power, Action and Belief,* edited by J. Law, 264–80. London: Routledge, 1986.
Lazlo, Ervin. *The Systems View of the World: A Holistic Vision for Our Time. Advances in Systems Theory, Complexity, and the Human Sciences.* Cresskill, NJ: Hampton, 1996.
Lazlo, Ervin, editor. "The Relevance of General Systems Theory." *Papers Presented to Ludwig von Bertalanffy on His Seventieth Birthday.* New York: Braziller, 1972.
Leaman, O. *Evil and Suffering in Jewish Philosophy.* Cambridge: Cambridge University Press, 1995.
Levins, Richard. "The Qualitative Analysis of Partially Specified Systems." *Mathematical Analysis of Fundamental Biological Phenomena: Annals of the New York Academy of Sciences* 231 (1974) 123–38.

Linnemann, Eta. *Gleichnisse Jesu: Einführung und Auslegung*. Göttingen: Vandenhoeck & Ruprecht, 1966.

Lipset, S.M. *American Exceptionalism*. New York: Norton, 1996.

Locker, Alfred. "Schöpfungs- und Evolutions-Problematik in system-theoretisch klassischer und transklassischer Sicht. Der Mensch im Widerspruch von Außen- und Innenbeobachtung sowie der Mitgestaltung von Ursprung und Ziel." In *Evolution im Diskurs. Grenzgespräch zwischen Naturwissenschaft, Philosophie und Theologie, Eichstätter Studien NF XXXIX*, edited by A. J. Bucher und D. S. Peters, 217-50. Regensburg: Pustet, 1998.

———. "'Evolution'- Ein faszinierender Ungedanke: Versuch und Misslingen einer Gestalt-Usurpation." *Zeitschrift für Ganzheitsforschung* Neue Folge 26 (1982) 17-39.

———. "'Synolologie'und 'Chaologie,' oder die widersprüchliche Einheit von Ganzheit, Gestalt und System. Vom Beobachten zum Schauen und wieder retour." In *Wege zur Ganzheit. Festschrift für J. Hanns Pichler zum 60. Geburtstag*, edited by G. E. Tichy, H. Matis, and F. Scheuch, 71-101. Berlin: Duncker & Humblot, 1996.

———. "'Grenzen des Wissens'—hauptsächlich menschengemacht." *Ethik und Sozialwissenschaft* 4.1 (1993) 48-51.

———. "Angriff auf die ganzheitliche Welt-Auffassung: Zurückweisung des Luhmannschen Ansatzes einer Systemtheorie. Ein erweiterter Besprechungsaufsatz." No pages: Unpublished paper, Vienna, 2003.

———. "Der Mensch im Angesicht suggestiver Verführung." *Österreichische Ärztegesellschaft* 42.18 (1978) 35-38.

———. "Der Mensch: Nicht unbeteiligter Zuschauer, sondern Mitgestalter am Weltgeschehen. Die Bedeutung von Meditation und Ekstase als Transklassche Mittel dazu." *Gnostika* 2 (1998) 34-42.

———. "Die Rolle des Beobachter-Subjekts in einer transklassischen Sicht der Geist/ Gehirn Problematik." *Ethik und Sozialwissenschaften, Streitforum für Erwägungskultur* 6 (1995) 89-101.

———. "Evolutionstheorie: Wissenschaft oder Ideologie." Manuscript of a lecture delivered on February 1986 to the Political Academy of the ÖVP.

———. "Evolutions-Theorie als Paradigma für endgültigen Wissenschaftsverfall?" *Ethik und Sozialwissenschaften. Streitforum für Erwägungskultur* 5 (1994) 228-31.

———. "Horizontale und vertikale Relationalität des Menschen: Differenz und Einheit jenseits der Beobachter-Perspektive." In *Die Biopsychosoziale Einheit Mensch— Begegnungen: Festschrift für K.-Fr. Wessel*, edited by F. Kleinhempel, A. Möbius, H.-U. Soschinka, and M. Waßermann, 408-16. Bielefeld: Kleine, 1996.

———. "Kybernetik und Systemtheorie als metatheoretische Brücken zwischen Einzelwissenschaften und Philosophie." In *Kybernetik und Systemtheorie. Wissensgebiete der Zukunft*, edited by E. V. Goldammer, H. Spranger, S. Fuchs, 23-43. Wessels: Greven, 1991.

———. "Leib—'höchste der Hieroglyphen'. Elemente einer synoptischen Somatologie." *Archiv für Religionspsychologie* 20 (1992) 194-218.

———. "Metatheoretical Presuppositions for Autopoiesis. Self-Reference and 'Autopoiesis.'" In *Autopoiesis: A Theory of Living Organization*. Vol. 3, edited by M. Zeleny, 209-33. New York: North Holland, 1981.

———. "On The Ontological Foundations of the Theory of Systems." In *Unity Through Diversity: A Festschrift for Ludwig von Bertalanffy*, edited by W. Gray and N. D. Rizzo, 537-57. London: Gordon & Breach, 1973.

———. "On the Origin of Systems and the Role of Freedom Therein." In *Improving the Human Condition. Quality and Stability in Social System. Proc. 25th Anniversary SGRS-Meeting*, edited by R. F. Ericson, 95–103. New York: Springer, 1979.

———. "Recent Approach to Trans-Classial Systems Theory. The Paradoxical Unity of Science with Non- and Super-Science." In *Advances in Systems Res. & Cybernetics*. Vol. III, edited by. G. E. Lasker, 11–16. Windsor, ON: IIAS, 1998.

———. "Selbstentstehung von Leben und Vernunft-ein Trugschluss. Die Unhaltbarkeit von Genesemodellen." In *Überlieferung und Aufgabe: Festschrift für Erich Heintel zum 70. Geburtstag*. Vol 2, edited by H. Nagl-Dolecal, 33–69. Vienna: Wilhelm Braumueller, 1981.

———. "The Autological Foundation and Actualization of Peace: The Role of the Observer and the Designer in the Peace Paradox." In *Advances in Sociocybernetics and Human Development IV*, edited by G. E. Lasker, 1–11. Windsor, ON: IIAS, 1998.

———. "The Healing of Mankind's Predicaments Through Suffering: A Paradoxical View, Based on Transclassical Systems Theory." In *Life and Healing: Healing of Nature, Healing Our Civilization and Healing Humankind. Proceeding*. Vol. 2. Third Yoko Civilization International Conference, 131–52. Tokyo: Yoko, 2001.

———. "The Present Status of Classical Systems Theory, 25 Years after Ludwig von Bertalanffy's Decease." In *Advances in Artificial Intelligence and Engineering Cybernetics*. Vol. V, edited by G. E. Lasker, 8–16. Windsor, ON: IIAS, 1999.

———. "Tierversuchs-Ethik und der ‚Menschenversuch.' Gedanken zum Umgang mit dem Tier." *Altex* 21 (2004) 221–26.

———. "Über Entstehung und Entwicklung formaler Systeme." *Nova Acta Leopoldina* Neue Folge 42.218 (1975) 498–503.

———. "Vorstoß zu einer transklassichen Sicht. Eine Betrachtung über den Zustand der Allgemeinen System-Theorie 25 Jahre nach Ludwig von Bertalanffys Tod." *Newsletter der Deutschen Gesellschaft für Systemforschung* 8 (1998/99) 8–20.

Locker, Alfred, ed. *Biogenesis, Evolution, Homeostasis: A Symposium by Correspondence*. Berlin: Springer, 1973.

———. ed. *Evolution kritisch gesehen*. Salzburg: Pustet, 1983.

Locker, Markus, and Clemens Sedmak. "The Language Game of Revelation: Interpreting the Book of Revelation through Wittgenstein's Philosophy of Language." *Philosophy & Theology* 13 (2002) 241–62.

———. "[il y a] or The Withdrawal to the Portal of Being. A systems-theoretical Appraisal of Emmanuel Levinas' 'path out of being." MA Thesis. Ateneo de Manila University, 2006.

———. "A.I. and Ethics: A Language Philosophical Question & Systems Theoretical Reply." In *Cognitive, Emotive and Ethical Aspects of Decision Making in Humans and Artificial Intelligence*. Vol. III, edited by Iva Smit, et al., 63–68. Windsor, ON: IIAS, 2004.

———. "Das Buch der Offenbarung im Verständnis der Sprachphilosophie." *Zeitschrift für Ganzheitsforschung* 46.3 (2002) 3–11.

———. "Glimpses of Truth: An Obituary for Alfred Locker." *Cybernetics and Human Knowing* 12.3 (2005) 103–5.

———. "Obituary Alfred Locker." *Systems Research and Behavioral Science* 22.6 (2005) 571–75.

———. "Reviving Paradoxes: Transclassical Systems Theory as Meta-theory for a Science-Faith Dialogue." E-conference proceedings: Continuity and Change 2006. Philadelphia, June 2006. Online: http://www.metanexus.net/conferences/pdf/conference2006/Locker.pdf.

———. "Seeing Actions of Hope in a World of Tragedy: Re-reading Matthew's Parable of the Talents." *Biblische Zeitschrift* 49 (2005) 161–73.

———. "Systems Theory and the Conundrum of ens: Thoughts and Aphorisms." *Foundations of Science* 11 (2006) 297–317.

———. "Systems-Theoretical Considerations on the Role of the Observer in Teilhard's Human Phenomenon: Viewing the Universe from Within." E-conference proceedings: Teilhard_Asia_2006. Manila, August 1 to September 31, 2006, 7 pages. Online: http://www.geochris.net/tasiapapers.htm.

———. "System-Theorie als Theorienbasis für die Theologie: Ein Versuch und Vorschlag." *IBW Journal* 4 (2004) 16–22.

———. "Offenbarung in Ganzheitlich-Systemtheoretischer Deutung." *Zeitschrift für Ganzheitsforschung* 46.1 (2002) 3–10.

Longenecker, R. N., ed. *The Road from Damascus: The Impact of Paul's Conversion on his Life, Thought, and Ministry*. Grand Rapids: Eerdmans, 1997.

Lorenz, E. N. "Deterministic Nonperiodic Flow." *Journal of Atmospheric Science* 20 (1963) 130–41.

Luhmann, N. *Social Systems*. Stanford: Stanford University Press, 1995.

Lurker, Manfred. "Adam Kadmon." In *Dictionary of Gods and Goddesses, Angels and Demons*, edited by M. Lurker, 6. New York: Routlege & Kegan Paul, 1987.

Luz, Ulrich. *Das Evangelium nach Matthäus (Mt 18–25)*. Evangelisch-Katholischer Kommentar zum Neuen Testament 1.3. Stuttgart: Benziger/Neukirchener, 1997.

MacAndrew, Alec. "Life: Puppetry or Pageantry? A Response to Cardinal Schönborn's attack on science." Online: http://www.evolutionpages.com/Schoenborn_critique.htm

Madsen, D. L. *American Exceptionalism*. Edinburgh: Edinburgh University Press, 1998.

Margalit, Avishai. *The Decent Society*. Harvard: Harvard University Press, 1996.

Marginson, S., P. Murphy, and M. A. Peters. *Global Creation: Space, Mobility and Synchrony in the Age of the Knowledge Economy*. New York: Lang, 2010.

Marion, Jean-Luc. *The Erotic Phenomenon*. Translated by Stephen E. Lewis. Chicago: University of Chicago Press, 2006.

Matt, Daniel C. "Ayin: The Concept of Nothingness in Jewish Mysticism." In *The Problem of Pure Consciousness*, edited by Robert K. C. Forman, 121–59. New York: Oxford University Press, 1990.

Maturana, H., and F. Varela. *Autopoiesis and Cognition: The Realization of the Living*. Boston: Reidel, 1980.

———. *The Tree of Knowledge: The Biological Roots of Human Understanding*. Boston: Random, 1987.

Maturana, H. "Biology of Language: The Epistemology of Reality." In *Psychology and Biology of Language and Thought: Essays in Honor of Eric Lenneberg*, edited by George A. Miller and Elizabeth Lenneberg, 27–63. New York: Academic, 1978.

———. *Biology of Cognition*. Urbana: University of Illinois, 1970.

McKeon, R. P. *Freedom and History and Other Essays*. Chicago: University of Chicago Press, 1990.

McKeon, R. P. *Selected Writings of Richard McKeon*. Chicago: University of Chicago Press, 1998.
McLuhan, M. *The Gutenberg Galaxy*. Toronto: University of Toronto Press, 1962.
McNeill, W. *Europe's Steppe Frontier, 1500–1800*. Chicago: University of Chicago Press, 1964.
———. *The Metamorphosis of Greece since World War II*. Chicago: University of Chicago Press, 1978
———. *Venice: The Hinge of Europe, 1081–1797*. Chicago: University of Chicago Press, 1974.
McTaggart-Ellis, J. "The unreality of time." *Mind* 68 (1908) 457–74.
Meurer, Herman-Joseph. *Die Gleichnisse Jesu als Metaphern, Bonner Biblische Beiträge 111*. Bonn: Philo, 1997.
Miller, Kenneth R. "The Cardinal's Big Mistake: Darwin Didn't Contradict God." *The Providence Journal* 10 August 2005. Online: http://www.millerandlevine.com/km/evol/catholic/projo.html.
Muck, O. *Rationalität und Weltanschauung*. Innsbruck: Tyrolia, 1999.
Müller, K. *Allgemeine Systemtheorie. Geschichte, Methodologie und Sozialwissenschaftliche Heuristik eines Wissenschaftsprogramms*. Opladen: Westdtsch, 1996.
Mumford, L. *The City in History*. Harmondsworth: Penguin, 1961.
Murphy, George L. "Does the Trinity Play Dice?" *Perspectives on Science and Christian Faith* 51 (March 1999) 18–25.
Murphy, P. "Architectonics." In *Agon, Logos, Polis*, edited by J. Arnason and P. Murphy, 207–32. Stuttgart: Steiner, 2001.
———. "Communication and Self-Organization." *Southern Review* 37.3 (2005) 87–102.
———. "France's Mediterranean Antipodes." In *The Mediterranean Reconsidered: A Multidisciplinary View*, edited by R. Hadj-Moussa, F. Loriggio, and M. Peressini, 249–60. Gatineau, Québec: Canadian Museum of Civilization, 2005.
———. "Marine Reason." *Thesis Eleven* 67 (2001) 11–38.
———. "The Ethics of Distance." *Budhi: A Journal of Culture and Ideas* VI 2.3 (2003) 1–24.
———. *Civic Justice: From Greek Antiquity to the Modern World*. Amherst, NY: Humanity, 2001.
———. "The Pitch Black Night of Human Creation: Calling Heidegger's Philosophy of Terror to the Account." In *Heidegger and the Aesthetics of Living*, edited by Vrasidias Karalis, 65–78. Newcastle: Cambridge Scholars, 2008.
Murphy, P., and Roberts, D. *Dialectic of Romanticism: A Critique of Modernism*. London: Continuum, 2004.
Needham, Joseph. *Science and Civilization in China*, vol. 2. History of Scientific Thought. Cambridge: Cambridge University Press, 1956.
Neiman, Susan. *Evil in Modern Thought: An Alternative History of Philosophy*. Princeton: Princeton University Press, 2002.
Nicolis, G, and I. Prigogine. *Self-Organization in Non-Equilibrium Systems*. New York: Wiley, 1977.
Nietzsche, Friedrich. *Die Geburt der Tragödie aus dem Geiste der Musik* (1872). Oxford: Oxford University Press, 2008.
Noble, D.W. *The End of American History*. Minneapolis: University of Minnesota Press, 1985.
Novak, M. *The Spirit of Democratic Capitalism*. New York: Simon & Schuster, 1982.

Novalis. *Die Christenheit oder Europa* (1799). *Die Werke Friedrich von Hardenbergs.* Vol. 3, 507–25. Stuttgart: Kohlhammer, 1960–1977.
Nozick, Robert. *The Examined Life.* New York: Touchstone, 1989.
Nussbaum, M. *Upheaval of Thought: The Intelligence of Emotions.* New York: Cambridge University Press, 2001.
Parpola, Simo. "The Assyrian Tree of Life: Tracing the Origins of Jewish Monotheism and Greek Philosophy." *Journal Near Eastern Studies* 52 (1993) 161–208.
Parsons, T., and E. Shils. *Towards a General Theory of Action.* Cambridge: Harvard University Press, 1951.
Parsons, T. *Societies: Evolutionary and Comparative Perspectives.* Englewood Cliffs, NJ: Prentice-Hall, 1966.
———. *The Social System.* New York: Free Press, 1951.
———. *The Structure of Social Action.* 2nd ed. New York: Free Press, 1949.
Patai, Raphael. *The Jewish Alchemists.* Princeton: Princeton University Press, 1994.
Patt-Shamir, Galia. *To Broaden the Way: A Confucian-Jewish Dialogue.* New York: Rowman & Littlefield, 2006.
Perec, G. *Life: A User's Manual.* London: Vintage, 2003.
Pichler, Josef H. "Alfred Locker im Gedenken." *Gnostika* (2006) 95.
Piepmeier, R. *Aporien des Lebensbegriffs seit Oetinger.* Freiburg: Alber, 1978.
Plaskow, J. "Facing the Ambiguity of God." *Tikkun* 6.5 (1991) 70–71.
Polanyi, Michael. *Personal Knowledge.* New York: Harper & Row, 1964.
Pollak, Michael. *Mandarins, Jews, and Missionaries: The Jewish Experience in the Chinese Empire.* Philadelphia: Jewish Publication Society of America, 1980.
Prigogine, I., and I. Strengers. *Order out of Chaos: Man's New Dialogue with Nature.* London: Bantam, 1985.
Rawrer, Karl, and Karl Rahner. "Weltall-Erde-Mensch." In *Christlicher Glaube in moderner Gesellschaft.* Vol. 3, edited by Franz Böckle, et al., 6–85. Freiburg: Herder, 1981.
Rawson, Philip. *The Art of Tantra.* Greenwich, CT: New York Graphic Society, Ltd., 1973.
Regardie, Israel. *A Garden of Pomegranates: An Outline of the Qabalah.* Saint Paul, MI: Llewellyn, 1970.
Ricoeur, Paul, and Eberhard Jüngel. *Metapher: Zur Hermeneutik religiöser Sprache. Sonderheft Evangelische Theologie.* München: Kaiser, 1974.
Rohrbaugh, Richard L. "A Peasant Reading of the Parable of the Talents/Pounds: A Text of Terror?" *Biblical Theological Bulletin* 23 (1993) 32–39.
Rorty, R. *Achieving Our Country.* Cambridge: Harvard University Press, 1998.
Rössler, Otto E. "Einstein Completion of Quantum Mechanics Made Falsifiable." In *Complexity, Entropy and the Physics of Information,* edited by W. H. Zurek, 367–73. Redwood City, CA: Addison-Wesley, 1990.
———. "Jumping Identities of Particles." *Symmetry: Culture & Science* 7 (1996) 307–19.
———. "Nonlinear Dynamics, Artificial Cognition and Galactic Export." In *Computing Anticipatory Systems,* edited by D. Dubois, 47–67. Melville, NY: American Institute of Physics, 2004.
———. "Relative-state Theory: Four New Aspects." *Chaos, Solitons & Fractals* 7 (1996) 845–52.
———. *Das Flammenschwert.* Berne: Benteli, 1996
———. *Endophysics: The World as an Interface.* Singapore: World Scientific, 1998.

———. "On the Animal-Man Problem from the Viewpoint of the Theoretical Biology of Behavior." *Schweizer Rundschau* 67 (1968) 529–32.

———. "Endophysics: Descartes Taken Seriously." In *Inside vs. Outside*, edited by H. Atmanspacher and G. J. Dalenoort, 153–61. Berlin: Springer, 1994.

Rössler, O. E., R. Rössler, and P. Weibel. "Is Physics an Observer-Private Phenomenon Like Consciousness?" *Journal of Consciousness Studies* 5 (1998) 443–53.

Roszak, Theodore. *Where the Wasteland Ends*. New York: Anchor, 1973.

Rothenberg, Albert. "The Process of Janusian Thinking in Creatiity." *Archives of General Psychiatry* 24.3 (1971) 195–205.

Ryan, James A. "Leibniz's Binary System and Shao Yong's Yijing." *Philosophy East and West* 46 (1996) 59–90.

Sacred Congregation of the Holy Office. "Monitum Concerning the Writings of Fr. Teilhard de Chardin." June 30, 1962, reiterated on July 20, 1981. Online: http://www.ourladyswarriors.org/dissent/cdfchard.htm.

Schick, Theodore, Jr. "Can Science Prove that God Does Not Exist?" *Free Inquiry Magazine* 21.1. Online: http://www.secularhumanism.org/library/fi/schick_21_1.html.

Schimmel, S. *Wounds Not Healed by Time: The Power of Repentance and Forgiveness*. New York: Oxford University Press, 2002.

Scholem, Gershom. *Major Trends in Jewish Mysticism*. New York: Schocken, 1961.

———. *New American Library*. New York: Meridian, 1974.

———. *On the Kabbalah and Its Symbolism*. New York: Schocken, 1969.

———. *On the Mystical Shape of the Godhead*. New York: Schocken, 1991.

———. *On the Possibility of Jewish Mysticism in Our Time & Other Essays*. Jerusalem: Jewish Publication Society, 1997.

———. *Origins of the Kabbalah*. Princeton: The Jewish Publication Society & Princeton University Press, 1987.

Schönborn, Christoph. *Ziel oder Zufall? Schöpfung und Evolution aus der Sicht eines vernünftigen Glaubens*. Vienna: Herder, 2007.

———. "Fides, Ratio, Scientia. Zur Evolutionismusdebatte." A lecture of Cardinal Christoph Schönborn during an alumni gathering in Castel Gandolfo, Sept. 1–3, 2006. Online: http://stephanscom.at/edw/reden/0/articles/2006/10/17/a11644/.

———. "Finding Design in Nature." New York Times, 7 July 2005. No pages. Online: http://www.millerandlevine.com/km/evol/catholic/schonborn-NYTimes.html.

Schöppe, Arno. *"Theorie paradox." Kreativität als systemische Herausforderung*. Heidelberg: Carl Auer, 1995.

Schottroff, Luise. "Human Solidarity and the Goodness of God: The Parable of the Workers in the Vineyard." In *God of the Lowly: Socio-Historical Interpretations of the Bible*, edited by W. Schottroff and W. Stegemann. Translated by M. J. O'Connell, 129–47. Maryknoll, NY: Orbis, 1984.

Schutz, A. *The Phenomenology of the Social World*. Evanston, IL: Northwestern University Press, 1967.

Sedmak, Clemens. "Die Frage: Können Religionen falsifziert werden?" In *Was wir Karl R. Popper und seiner Philosophie verdanken*, edited by E. Morscher, 317–48. St. Augustin: Academia, 2002. 317–48.

———. *Lokale Theologien und globale Kirche*. Freiburg: Herder, 2000.

———. *Theologie in nachtheologischer Zeit*. Mainz: Gruenewald, 2003.

Shannon, C., and W. Weaver. *The Mathematical Theory of Communication*. Urbana: University of Illinois Press, 1949.

Simon, Herbert. *The Sciences of the Artificial*. Cambridge, MA: MIT, 1996.
Simon, Y. R. *Philosophy of Democratic Government*. Chicago: University of Chicago Press, 1951.
Sloterdijk, P. *Sphären*. Vols. I–III. Frankfurt: Suhrkamp, 1989/1999/2004.
Smit, Wim. "Free Will and Determinism." Unpublished paper. Baden-Baden: IIAS, 2005.
Spämann, Robert, ed. *Evolutionstheorie und menschliches Selbstverständnis. Zur Philosophischen Kritik eines Paradigmas moderner Wissenschaft*. CIVITAS Resultate. Vol. 6. Weinheim: Acta humanoria, 1984.
Spence, Jonathan. "A Leaky Boat to China." *New York Times Book Review* (19 October 1997). Online: http://www.nytimes.com/books/97/10/19/reviews/971019.19spencet.html.
Spencer Brown, George. *Laws of Form*. New York: Julian Press, 1972.
Spengler, O. *The Decline of the West*. New York: Knopf, 1926–28.
Spitz, R. A. "Hospitalism: An Inquiry into the Genesis of Psychiatric Conditions in Early Childhood." *The Psychoanalytic Study of the Child* 1 (1945) 53–74.
Strauss, L. "Jerusalem and Athens: Some Preliminary Reflections" in *Faith and Political Philosophy: The Correspondence Between Leo Strauss and Eric Voegelin 1934–1964*, edited by P. Emberley and B. Cooper, 197–208. University Park: Pennsylvania State University Press, 1993.
———. "The Three Waves of Modernity" In *Introduction to Political Philosophy*, 81–98. Detroit: Wayne State University Press, 1989.
———. *Natural Right and History*. Chicago: University of Chicago Press, 1953.
Taux, Ernst. "Die Verwendung erkenntnistheoretischer Begriffe in der theoretischen Biologie Uexkülls und Bertalanffys." In *Wissenschaft der Wendezeit-Systemtheorie der Alternative*, edited by J.-P. Schramm and E. Schramm, 83–88. Frankfurt: Fischer, 1988.
Thomas Aquinas. *Summa Theologiae*. New York: Benziger, 1947.
Ticciati, S. *Job and the Disruption of Identity: Reading beyond Barth*. London: T. & T. Clark, 2005.
Tishby, Isaiah. *The Wisdom of the Zohar*. Vol. 1. Translated by D. Goldstein. Oxford: Oxford University Press, 1989.
Townes, Charles H. "Logic and Uncertainties in Science and Religion." In *Science and Theology: The New Consonance*, edited by T. Peters, 43–55. Boulder, Colorado: Westview, 1998.
Toynbee, A. *A Study of History*. London: Oxford University Press, 1934.
Turner, V. *The Ritual Process*. Chicago: Chicago University Press, 1969.
Vanier, Jean. *Becoming Human*. New York, Paulist, 1987.
Varela, F. *Principles of Biological Autonomy*. New York: Elsevier North Holland, 1979.
Via, Dan Otto. *The Parables: Their Literary and Existential Dimensions*. 1967. Reprint, Eugene, OR: Wipf & Stock, 2007.
Voegelin, E. *On the Form of the American Mind*. Baton Rouge: Louisiana State University Press, 1955.
———. *Order and History*. Vols. 1–5. Baton Rouge: Louisiana State University Press, 1956–87.
———. *The Collected Works of Eric Voegelin*. Baton Rouge: Louisiana State University Press, 1989.
———. *The New Science of Politics*. Chicago: University of Chicago Press, 1952.

Wang, Robin R. "Zhou Dunyi's Diagram of the Supreme Ultimate Explained (Taijitu shuo): A Construction of the Confucian Metaphysics." *Journal of the History of Ideas* 66.3 (2005) 307–23.

Weber, D. "From Limen to Border: A Meditation on the Legacy of Victor Turner for American Cultural Studies." *American Quarterly* 473 (1995) 525–36.

Weeks, W. E. *Building the Continental Empire*. Chicago: Dee, 1996.

Weinberg, G. M. *An Introduction to General Systems Thinking*. New York: Wiley, 1975.

———. *Rethinking Systems Analysis and Decision*. New York: Dorset, 1988.

Welch, Holmes. *Taoism: The Parting of the Ways*. Boston: Beacon, 1971.

Wiener, N. *Cybernetics: or Control and Communication in the Animal and the Machine*. Cambridge: MIT, 1985.

Wilhelm, Richard. *The Secret of the Golden Flower*. New York: Harcourt, Brace & World, 1962.

Williams, W. A. *The Roots of the Modern American Empire*. New York: Random, 1969.

———. *The Tragedy of American Diplomacy*. Cleveland: World, 1959.

Wilson, B. *Systems: Concepts, Methodologies, and Applications*. Chichester, UK: Wiley, 1984.

Wing-tsit Chan, editor. *Wing-tsit Chan, Chu Hsi and Neo-Confucianism*. Honolulu: University of Hawaii Press, 1986.

Wittgenstein, Ludwig. *Philosophical Investigations*, edited by G. E. M. Anscombe and R. Rhees. Oxford: Blackwell, 1953.

———. *On Certainty*, edited by G. E. M. Anscombe and G. H. von Wright. New York: Harper & Row, 1972.

Wollheim, R. *On the Emotions: Ernst Cassirer Lectures*. New Haven: Yale University Press, 1999.

Wolters, G. "Modell." In *Enzyklopädie Philosophie und Wissenschaftstheorie*. Vol. 2, edited by J. Mittelstraß, 911–92. Mannheim: Metzler, 1984.

Yates, E. ed. *Self-Organizing Systems: the Emergence of Order*. New York: Plenum, 1987.

Yudelove, Eric. *The Tao & The Tree of Life: Alchemical & Sexual Mysteries of the East and West*. St. Paul, MN: Llewellyn, 1995.

Zeleny, M., ed. *Autopoiesis: A Theory of Living Organization*. New York: North Holland, 1981.

Zwick, Martin. "Overview of Reconstructability Analysis." *Kybernetes* 33 (2004) 877–905.

———. "The Diagram of the Supreme Pole and the Kabbalistic Tree: On the Similarity of Two Symbolic Structures." *Religion East & West: the Journal of the Institute for World Religions* 9 (Oct 2009) 76–87.

Contributors

DAVID J. Krieger, *Privatdozent* at the University of Lucerne, was awarded habilitations in the Science of Religions and Communication Science. He is the co-founder of the Institute for Communication and Culture at the University of Lucerne, and since 2006 co-director of the Institute for Communication Research in Lucerne. Among his numerous publications are the books: *The New Universalism. Foundations for a Global Theology* (1991); *Einführung in die Allgemeine Systemtherorie* (1996); *Kommunikationssystem Kunst* (1997), and co-authored with A. Belliger, *Ein einführendes Handbuch zur Akteur-Netzwerk-Theorie* (2006).

ALFRED LOCKER (d. 2005) held the position of Professor for Theoretical Physics at the Vienna University of Technology. His research spanned from the critique of evolutionary theory to systems theory and systems related issues in philosophy. He edited the volumes: *Evolution, kritisch gesehen* (1983) and *Biogenesis, Evolution, Homeostasis: A Symposium by Correspondence* (1993), and authored more than 100 articles, among them: "Hamann und die Naturwissenschaft von heute. Vorweggenommene Kritik der Allgemeinen System-Theorie" (2005), and "Karl Rahner—Sprachvertuschte Zerstörung der Theologie" (2004).

MARKUS LOCKER is Associate Professor for Theology at the Ateneo de Manila University and currently completing a second PhD at Monash University. His research focuses on biblical theology, systems epistemology, and science and faith communications. He has published *The New World of Jesus' Parables* (2008); edited *Led By the Spirit* (2002) and authored "Systems Theory and the Conundrum of *ens*: Thoughts and Aphorisms" (2006); "Reading and Re-reading Matthew's Parable of the Talents in Context" (2005), and "'Jesus' Language-Games: The Significance of the Notion of Language-Game for a Reformulation of 'New Testament Biblical Theology" (2009).

OTTO E. RÖSSLER is Professor for Theoretical Biochemistry at the University of Tübingen. Designing the well-known Rössler attractor in 1976 he recently prompted the critical debate on the possible consequences of operating the Large Hadron Collider. His more than 340 scientific publications stretch from topics of biogenesis to chaos research and endopysics. Prof. Rössler has authored ten books, among them: *Encounter with Chaos* (1992); *Endophysics: The world as an interface* (1992); *Das Flammneschwert* (1996); together with Reimara Rössler, *Das Denken eines Kindes. Entwicklung, Persönlichkeit, Gefühle* (1998), and with Artur P. Schmidt, *Medium des Wissens. Das Menschenrecht auf Information* (2000).

MARTIN ZWICK is Professor of System Science at Portland State University. His main research areas are information theoretic modeling, machine learning, theoretical biology, game theory, and systems theory and philosophy. He has extensively published on discrete multivariate modeling (reconstructability analysis), "artificial life"/theoretical biology, and systems philosophy. Among his scientific publications are: "Control Uniqueness In Reconstructability Analysis" (1996); "Spinoza and Gödel: Cuasa Sui and Undecidable Truth" (2007), and together with Jeffrey A. Fletcher, "Strong Altruism Can Evolve in Randomly Formed Groups" (2004).

CLEMENS SEDMAK earned three doctorates and two habilitations, and currently holds the F. D. Maurice Professorial Chair of Moral and Social Theology at King's College, London. He is concurrently Professor for Epistemology at the University of Salzburg and director of the Center for Ethics and Poverty Research at the same university. He works in the areas of social ethics, epistemology, and the philosophy of religion, religious studies, and poverty research. Prof. Sedmak has edited ten volumes and authored more than twenty books. Among his most recent publications are: *Erkennen und Verstehen, Grundkurs Erkenntnistheorie und Hermeneutik* (2003); *Katholisches Lehramt und Philosophie. Eine Verhältnisbestimmung* (2003); *Die Politische Kraft der Liebe* (2007), and *Religious Intelligence: Developing Religious Literacy in a Secular World* (2009).

PETER MURPHY is Associate Professor of Communications and Director of the Social Aesthetics Research Unit at Monash University. He works in the areas of social theory and philosophy, political economy, aesthetics and communications. He is the author of *Civic Justice: From Greek Antiquity to the Modern World* (2001), the co-author with David Roberts of *Dialectic of Romanticism: A Critique of Modernism* (2004); and, with Michael Peters and Simon Marginson, of *Creativity and Global Knowledge Economy* (2009), *Global Creation* (2010), and *Imagination* (2010). Murphy's body of work also includes more than eighty journal articles and chapters in edited collections.

www.ingramcontent.com/pod-product-compliance
Lightning Source LLC
Chambersburg PA
CBHW051639230426
43669CB00013B/2370